Lecture Notes in Business Information Processing 435

More information about this series at http://www.springer.com/series/7911

Adiel Teixeira de Almeida ·
Danielle Costa Morais (Eds.)

Innovation for Systems Information and Decision

Third Innovation for Systems Information and Decision Meeting, INSID 2021 Virtual Event, December 1–3, 2021 Proceedings

Editors
Adiel Teixeira de Almeida
Universidade Federal de Pernambuco
Recife, Brazil

Danielle Costa Morais
Universidade Federal de Pernambuco
Recife, Brazil

ISSN 1865-1348 ISSN 1865-1356 (electronic)
Lecture Notes in Business Information Processing
ISBN 978-3-030-91767-8 ISBN 978-3-030-91768-5 (eBook)
https://doi.org/10.1007/978-3-030-91768-5

This Springer imprint is published by the registered company Springer Nature Switzerland AG
The registered company address is: Gewerbestrasse 11, 6330 Cham, Switzerland

Preface

The INnovation for Systems Information and Decision (INSID) meeting is an event (http://insid.events) linked to the international network INCT-INSID (http://insid.org.br). This network involves academics and practitioners from different countries, bringing together outstanding researchers from around the world in the field of information systems and decision.

The INSID meetings have provided a stimulating environment for the dissemination of state-of-the-art thinking and knowledge about INnovation for Systems, Information and Decision. This broad theme is transversely related to many areas, particularly to operational research, management engineering (or production engineering), including also systems engineering (and engineering in general), management science, computer science, and their interdisciplinary related areas. These meetings have prompted discussions among participants and the exchange of ideas and critical comments for further improvement since 2008, under the acronym SIDS.

INSID 2021 was to have been held at the Federal University of Pernambuco, in Recife-Pernambuco, Brazil, during December 1–3, 2021. However, due to the COVID-19 pandemic, it took place virtually (as did INSID 2020). Thus, this was the second time that the event took place under an online format. Moreover, this is the second volume of INSID meetings in the Lecture Notes in Business Information Processing (LNBIP) series.

In total, 70 papers were approved for presentation covering the main topics related to the themes and areas of interest of the meeting as follows: methodological advances in decision-making and aid; decision models in the environmental context; decision models in the energy context; decision models in service systems; and potential applications of decision and negotiation models. After a thorough review process, nine of these papers were selected for inclusion in this volume of INnovation for Systems Information and Decision: Models and Applications.

These nine papers reflect methodological improvements and advances in Multicriteria Decision-Making/Multicriteria Decision-Aid (MCDM/MCDA) oriented toward real-world applications, which contribute to the understanding of relevant developments of current research on and future trends of INnovation for Systems Information and Decision.

The first paper by Czekajski et al. develops an application of the FITradeoff method to identify the potential of the cultural heritage of the Czeladź commune and to use it to analyze a possible set of Cultural Tourism Products (CTPs). To do so, they take a formal multicriteria decision-aiding approach. The second paper by Danielson and Ekenberg presents a review of some leading algorithms for automatic weight generation without external parameters besides cardinal and ordinal rankings and provides some guidelines for selecting a surrogate weight-generating function for MCDM applications, in ordinal as well as cardinal information settings. The third paper by Wan et al. combines the Data Envelopment Analysis (DEA) model and a technique for order performance by similarity to ideal solution (TOPSIS) to evaluate and then rank the efficiency and competitiveness of a Medium-lift Launch Vehicle (MLV).

Rai et al. deal with the analysis of a strategic port alliance in Japan based on cooperative game theory, by examining the International Container Strategy started in 2011 and the designation of Strategic International Ports. Espirito Santo at al. propose an improvement to the intra-criteria evaluation step of the FITradeoff method, by putting forward a new approach for eliciting marginal value functions based on partial information. It follows a study by Mondadori et al. which presents the use of the Multicriteria Partial Information Method for choosing the most suitable online platform to integrate hardware and consulting services for online data acquisition and a manufacturing execution system.

Vieira et al. propose an approach for solving multicriteria decision-making problems with hierarchically structured criteria in the FITradeoff method for choice and ranking problematics. Cimadamore et al. present an innovative approach to conduct pairwise comparisons for AHP based on a UI widget that resembles an interactive data plot. Finally, the ninth paper, by Syrides et al., presents a multimethodology for structuring and proposing interventions called Complex Holographic Assessment of Paradoxical Problems (CHAP^2) to support a post-graduation course on implant dentistry.

The preparation of this volume required the efforts and collaboration of many people. In particular, we would like to thank the Steering Committee and Program Committee for their contributions to INSID 2021. Special thanks also go to all members of the INCT-INSID network. We are also very grateful to the following reviewers for their timely and informative additional reviews: Marc Kilgour, Liping Fang, Pascale Zarate, Tomasz Wachowicz, Ana Paula Gusmão, Mischel Carmen Neyra Belderrain, Eduarda Frej, Leandro Rego, Maisa M. Silva, Carolina Lino, Jonatas de Almeida, Luciana Hazin, Ana Paula Cabral, and Alexandre Alberti.

We would also like to thank Ralf Gerstner, Alfred Hofmann, Christine Reiss, Guido Zosimo-Landolfo, and Anna Kramer at Springer for their excellent collaboration.

Finally, we hope readers will find the content of this book useful and stimulating and that it encourages them to seek to produce further developments and applications of INnovation for Systems Information and Decision.

December 2021

Adiel Teixeira de Almeida
Danielle Costa Morais

Organization

Program Chair

Danielle Costa Morais	Universidade Federal de Pernambuco, Brazil

Steering Committee

Adiel Teixeira de Almeida	Universidade Federal de Pernambuco, Brazil
Keith Hipel	University of Waterloo, Canada
Love Ekenberg	Stockholm University, Sweden
Marc Kilgour	Wilfrid Laurier University, Canada
Pascale Zarate	Université Toulouse 1 Capitole, France
Ralph Keeney	US Marketing and Decisions Group Inc., USA
Roman Slowinski	Poznan University of Technology, Poland
Rudolf Vetschera	University of Vienna, Austria
Petr Ekel	Pontitifica Universidade Católica de Minas Gerais, Brazil
Marcos Pereira Estellita Lins	Universidade Federal do Rio de Janeiro, Brazil
Helder Gomes Costa	Universidade Federal Fluminense, Brazil
Mischel Carmen Neyra Belderrain	Instituto Tecnologico de Aeronautica, Brazil
Danielle Costa Morais	Universidade Federal de Pernambuco, Brazil

Program Committee

Alexandre Bevilacquea Leoneti	Universidade de São Paulo, Brazil
Ana Paula Cabral Seixas Costa	Universidade Federal de Pernambuco, Brazil
Ana Paula Henriques de Gusmão	Universidade Federal de Sergipe, Brazil
Annibal Parracho Sant'Anna	Universidade Federal Fluminense, Brazil
Carlos Francisco Simões Gomes	Universidade Federal Fluminense, Brazil
Caroline Maria de Miranda Mota	Universidade Federal de Pernambuco, Brazil
Cristiano Alexandre V. Cavalcante	Universidade Federal de Pernambuco, Brazil
Cristiano Torezzan	Universidade Estadual de Campinas, Brazil
Daniel Aloise	Polytechnique Montréal, Canada
Haiyan Xu	Nanjing University of Aeronautics and Astronautics, China
Hannu Nurmi	University of Turku, Finland
João Carlos Correia Baptista Soares de Mello	Universidade Federal Fluminense, Brazil

Johannes Siebert	MCI Management Center Innsbruck, Austria
José Rui Figueira	Technical University of Lisbon, Portugal
Leandro Chaves Rêgo	Universidade Federal do Ceará, Brazil
Liping Fang	Ryerson University, Canada
Luciana Hazin Alencar	Universidade Federal de Pernambuco, Brazil
Luiz Bueno da Silva	Universidade Federal da Paraíba, Brazil
Luiz César Ribeiro Carpinetti	Universidade de São Paulo, São Carlos, Brazil
Maria Teresinha Arns Steiner	Pontifícia Universidade Católica Paraná, Brazil
Mariana Rodrigues de Almeida	Universidade Federal do Rio Grande do Norte, Brazil
Salvatore Greco	University of Catania, Italy
Sérgio Eduardo Gouvêa da Costa	Pontifícia Universidade Católica Paraná, Brazil
Vanessa Batista de Sousa Silva	Universidade Federal de Campina Grande, Brazil

Contents

Contents

FITradeoff Based Analysis of Cultural Tourism Products Regarding Post-industrial Heritage in Czeladź Commune in Poland

Marek Czekajski[1(✉)], Tomasz Wachowicz[1], and Eduarda Asfora Frej[2]

[1] Department of Operations Research, College of Informatics and Communication, University of Economics, 50, 1 Maja Street, 40-287 Katowice, Poland
marek.czekajski@edu.uekat.pl, tomasz.wachowicz@uekat.pl

[2] Center of Decision Systems and Information Development – CDSID, Universidade Federal de Pernambuco, Av. Da Arquitetura – Cidade Universitária, Recife, PE, Brazil
eafrej@cdsid.org.br

Abstract. Diverse cultural heritage, occurring on a local or regional scale, has become an extremely valuable base for creating new interesting tourist products. One example is the post-industrial heritage related to the areas of two former hard coal mines in the Czeladź Commune in the Silesian Voivodeship. The processes related to designing, planning and final decisions regarding selecting the best solution for the promotion of post-industrial heritage are, in this problem, non-trivial, as they have a complex structure and multiple criteria character. This paper presents specific characteristics of the decision problem under consideration resulting in several alternative options for a planned cultural tourism product (CTP) regarding post-industrial heritage in Czeladź. The analysis of its features, attributes, criteria and alternatives is provided to formulate a multiple criteria decision making (MCDM) problem. It is based on the 12-step analytic framework that implements the FITradeoff method for preference modelling. This method is operated by means of an interactive Decision Support System (DSS) with several flexibility features; moreover, the elicitation process is carried out based on partial information provided, which lead to time and effort saving.

Keywords: Multiple criteria analysis · Multiple criteria problems in cultural tourism · Decisions on new cultural tourism products · FITradeoff method

1 Introduction

In the 21st century, cultural tourism plays an important role for states or local authorities taking into account its social, economic and business consequences. Newer and newer regions or tourism products are involved in the international and domestic tourism trends. In the ever-growing competition, only such a tourism destination of tourism entity can survive that reveals the highest quality standards [23]. Cultural heritage is a cornerstone of local, regional, national and European identity. It should be handled with an integrated, complex approach when planning the regional development, as it is one of the most critical cultural, environmental and economic resources [12].

A. T. de Almeida and D. C. Morais (Eds.): INSID 2021, LNBIP 435, pp. 1–19, 2021.
https://doi.org/10.1007/978-3-030-91768-5_1

In economic and marketing terms, cultural tourism manifests itself in the existence of various products. Compared to the other types of products and services, tourism products have different characteristics, and there is a need to understand their specificity. The process of creating local and/or regional CTPs has a multifaceted nature. There are issues involved in creating CTPs, ranging from organisational, technical, infrastructural to economic and financial ones, such as management style, materials, devices, equipment and infrastructure, budget and the leading institutions. These resources may belong to different owners (local government authorities, different entities, institutions); hence various stakeholders may create CTPs. Besides these formal decision-makers, other parties may also be involved, such as advisers, experts, directors and managers of various institutions and entities (museum, community centres) whose scope of statutory activity is related to the management and development of cultural tourism issues. They may also have different goals and priorities; hence the evaluation of CTPs may require using many criteria.

Many studies apply decision analysis and appropriate multiple criteria decision aiding (MCDA) methods to analysing existing tourist products or planning new ones. For instance, a fuzzy multiple criteria decision making model (FMCDM) was used by Chou, Hsu, Chen [15] to select the location of an international tourist hotel in Taiwan. They created 21 criteria for selecting the hotel location and demonstrated the computational process and effectiveness of the fuzzy approach. Further, the Canadian governmental commission used several innovative approaches to collaborative decision making for environmental resources to the strategic spatial planning process for central British Columbia [47]. It (1) proposed to use these criteria to evaluate the effectiveness of the process from the perspective of tourism stakeholders (including cultural tourism) participating in the planning exercise, (2) made recommendations to governmental and tourism organisations on managing future spatial planning strategies to involve stakeholders in such processes. Finally, Wong and Fung [48] indicate the possibility of incorporating multi-criteria decision analysis procedures into ecotourism planning. They note that the geographic information system combined with the multiple criteria decision analysis procedure (GIS-based MCDA approach) can efficiently identify potential sites for various ecotourism activities and tourism development potential on Hong Kong's Lantau Island.

In the view of the above, it is clear that also the processes of creating post-industrial CTPs, should be viewed as multiple criteria management decisions. The following issues influence the complexity of the decision problem related to creating CTPs:

1) Different visions, goals and functions of CTPs presented by many stakeholders [38, 39, 45].
2) Many different evaluation criteria and various stakeholders' preferences, as well as the complexity of the data and information which need to be processed (e.g. a number of attributes that can be viewed as criteria; a number of different types of post-industrial heritage, a number of social, economic, technological factors, etc.) [24, 28, 29].
3) Problems of scales used to measure evaluation criteria – some criteria are quantitative, others are qualitative [16, 46].

4) Selection of the most important criteria that will best match the specificity and features of CTP related to the promotion of post-industrial heritage from among all possible criteria (examples of unified criteria are presented in Sect. 4.3) [8, 10, 27, 33, 42, 44].
5) Comparability of alternatives that can be of different type and form, e.g. product-thing, product-service, or product-route – which can take the real, virtual or mixed. [13, 29, 30, 43, 49].
6) Diversity of the post-industrial heritage base (what elements of this heritage should be used; how to classify these elements; what is the core of this heritage?).

The main objective of this paper is to identify the potential of the cultural heritage of Czeladź Commune and use it to analyse the possible set of CTPs implementing a formal MCDA approach. A decision-maker (DM) may use such analysis to enhance their knowledge of the problem and prepare for negotiations with other stakeholders when the local authorities consider the strategic plan of building such a product. A thorough historical literature review allows us to identify all possible remainings of the post-industrial heritage of Czeladź Commune that can be used to design complex CTPs. Then, by implementing the 12-step framework for structuring and analysing the MCDM problems [3], we structure the problem of ranking some selected instances of CTPs given the preselected set of the evaluation criteria. In this paper, the analysis is conducted from the viewpoint of the single stakeholder, representing the cultural institutions that may be involved in providing the services for such a future CTP, i.e. the "Saturn" Museum.

This paper is structured as follows. Section 2 identifies the post-industrial heritage of the Czeladź Commune that may be used to design the CTP. Section 3 presents how the 12-step framework for structuring and analysing the MCDM problem may be implemented to design the CTP related to local post-industrial heritage in Czeladź. Structuring the problem of post-industrial CTP design and choice is presented in Sect. 4. Section 5 analyses the preferences of the "Saturn" Museum managers, and Sect. 6 presents sensitivity analysis and recommendations. Some concluding remarks are presented in Sect. 7.

2 Post-industrial Heritage of the Czeladź Commune

2.1 The Heritage Identified

To identify the post-industrial heritage of the Czeladź Commune, we conducted a historical literature review. It allowed us to recognise the most important archival materials, such as (1) archival collection of the "Saturn" Museum in Czeladź [6, 7], (2) State Archive's resources regarding the Mining and Industrial Society "Saturn" [40], (3) State Archive's resources regarding the Nameless Society of Coal Mines "Czeladź" in Czeladź-Piaski [41]. These materials were supplemented by an additional, though surprisingly small, set of scientific manuscripts devoted to the industrial and historic issues of this part of Poland (see [9, 14, 17, 25, 26]). A thorough study of the sources mentioned above resulted in the identification of two former mines, i.e. "Saturn" and "Ernest-Michał" as being the main composites of the post-industrial heritage of Czeladź Commune. The heritage base of the former "Saturn" coal mine amounts to the following elements:

- Type I: Facilities (buildings, architectural objects) included in the "Saturn" coal mine.
- Type II: Housing estate – workers' and clerks' housing.
- Type III: Housing estate – public utility buildings (school, teachers' house, clerks' club and others).
- Type IV: Housing estate – management buildings.
- Type V: Machine and equipment infrastructure (e.g. "Wanda" power generator – reversible compressor, compressor by Belliss & Morcom, power systems – generators sets, control and measurement desk).
- Type VI: Parks, gardens and estate greenery.
- Type VII: Sports fields and other sports facilities.

In turn, the post-industrial heritage of the former "Ernest Michał" coal mine can be divided into the following several types:

- Type I: Housing estate – workers' and clerks' housing.
- Type II: Housing estate – public buildings (e.g. pharmacy, hospital, clerk's club, schools, an orphanage for children).
- Type III: Housing estate – management buildings.
- Type IV: Church, parishes (church buildings and other church infrastructure, e.g. parish cemetery, so-called "Catholic House", presbytery).
- Type V: Housing estate greenery.
- Type VI: Playing fields.
- Type VII: Objects included in the former coal mine, e.g., mine railway station, railway siding (connecting with the Warsaw-Vienna Iron Road), sorting building, central power plant, boiler room building, shaft structures and equipment.

The identified post-industrial heritage of the two aforementioned coal mines and their diverse infrastructure constitute a solid base and a promising potential for the creation of new cultural tourism products.

2.2 Considering Various CTPs to Be Built from the Heritage Identified

The problem under consideration is to analyse the quality of possible CTPs given the identified post-industrial heritage for Czeladź Commune. However, designing the potential products to be evaluated is not an easy task. The CTP may be a single homogeneous product (mono-product), but it may also consist of many different products, each having a different form and type. Many typologies are used in the literature to identify tourism products. From the viewpoint of possible products' categories, the following products can be distinguished [13, 29, 30, 43, 49]:

a) product-thing or product-material good (tourist guidebook, map, etc.),
b) product-service (thematic tourist guide, hotel and catering services, etc.),
c) product-event (post-industrial festivities, picnics, etc.),
d) product-services set (themed rallies, trips, running competitions, etc.),
e) product-object (museum, post-industrial monuments, buildings),
f) product-route (traditional, real post-industrial heritage route, etc.),

g) product-area (comprehensive post-industrial heritage area).

It is worth noting that a product-service is a single service related to tourist activities, e.g. transport, accommodation, catering, guide. In turn, the product as a set of services is at least two simple tourist services (trip, rallies, etc.)

CTPs can also be viewed in terms of the degree of integration of the aforementioned product categories [34]. This allows to identify of two types of products:

a) basic products, which are single products of minor complexity, such as a thing or a service;
b) integrated products (multi-products), with much greater complexity, organisational and/or spatial integration, which occurs on two levels:

- level I: basic product + organisation + management – CTP as an event, CTP as a tourist service set
- level II: basic product + organisation + management + location – CTP as an object (facility), CTP as a route, CTP as an area.

The third criterion for distinguishing CTPs is their form, namely:

1) real (traditional, material),
2) multimedia (e.g. related to the use of digital photos, films, or animations),
3) virtual (in the Internet as applications for PC and mobile devices),
4) mixed – hybrid (any combination of the forms given above).

In view of the above, the primary challenge is to create a product based on the potential provided by the post-industrial cultural heritage in the present Czeladź Commune, simultaneously integrating the diversity of CTP categories, types and forms. Various approaches may be used here to determine the set of alternative CTPs using some formal approaches as portfolio selection or the multidimensional knapsack models [2, 35]. Such a portfolio product consists of elements form the categories above of tourist products, their instances, and their forms of organisation (simple, complex products). Here, a problem arises regarding the appropriate selection of these elements in terms of quantity or compatibility. It lies, however, behind the scope of this paper.

3 12-Step Framework for Structuring and Analysing Multiple Criteria Problem

The evaluation of alternative variants of CTP related to post-industrial heritage is a decision making problem. Various methods developed by soft operations research may be used to precisely define, structure, and analyse it. Examples of such methods are the classic PrOACT (Problem, Objectives, Alternatives, Consequences, Tradeoffs) algorithm [22] or mind or cognitive mapping [18]. In this paper, we implement a more detailed algorithm that allows a thorough step-by-step analysis of the problem under consideration, i.e. the 12-step framework proposed by De Almeida et al. [3]. This framework includes the model building process, which has three main phases, each one with several steps as follows:

- First phase (Preliminary phase):
  - Step 1 – Characterising the Decision Maker
  - Step 2 – Identifying Objectives
  - Step 3 – Establishing Criteria
  - Step 4 – Establishing the Set of Actions and Problematic
  - Step 5 – Identifying the State of Nature
- Second phase (Preference Modelling and method choice):
  - Step 6 – Preference Modelling
  - Step 7 – Conducting and Intra-Criterion Evaluation
  - Step 8 – Conducting and Inter-Criterion Evaluation
- Third phase (Finalisation):
  - Step 9 – Evaluating Alternatives
  - Step 10 – Conducting a Sensitivity Analysis
  - Step 11 – Drawing up Recommendations
  - Step 12 – Implementing Actions

The first phase allows structuring the decision problem. The second one elicits the preferences from the decision-maker, while the third one – aggregates the preference information to provide the final evaluation of alternatives and produce the final recommendation on selecting the best (most preferred) one. The next three chapters discuss the steps of all these three phases when applied to designing and selecting CTP of post-industrial heritage for Czeladź Commune.

4 Structuring the Problem of Post-industrial CTP Design and Choice

4.1 Characterising Decision-Makers

In the first step, we characterise the decision-makers and other actors (stakeholders). Based on the analysis of human resources and the scope of competencies of various entities operating in the Czeladź Commune, it is possible to define a set of DMs and stakeholders who can participate in the process of creating a new CTP, namely:

1) Stakeholders at the level of the local government unit (LGU) of the Czeladź Commune:

a) formal DMs (direct, actual DMs with the power of decision): the Mayor of Czeladź Commune and his deputies,
b) substantively competent employees of organisational units (offices, departments) of the Czeladź Commune Hall, responsible for the promotion of local government units, territorial marketing, culture, tourism, city development, etc.

2) Stakeholders in units, entities and institutions subordinate to the Czeladź Commune as LGU, i.e. the DMs from:

a) municipal cultural institutions, e.g. "Saturn" Museum, "Kopalnia Kultury" ("Culture Mine") culture centre (directors, managers),
b) institutions related to tourism, sport, and recreation, e.g. MOSiR (Municipal Sports and Recreation Centre) (directors, managers).

3) Stakeholders in non-governmental organisations (NGOs) whose statutory activity is related to promoting culture, tourism, cultural tourism, e.g. managers in NGOs such as: Stowarzyszenie Edukacyjno-Kulturalne (Educational and Cultural Association) "Razem", Towarzystwo Powszechne (Public Society) "Czeladź".
4) Stakeholders in municipal educational institutions – schools – related to the subject of cultural tourism): teachers (e.g. in history, cultural studies, social studies, etc.), headmasters and/or their deputies.

We can also distinguish one more group of CTP stakeholders, namely tourists (culture tourism consumers) from both the Czeladź Commune and the Będzin District or the Silesian Voivodeship. However, at the level of analysis related to the structuring of the problem of creating CTP (concepts, models, various approaches), the scope of stakeholders has been focused on entities that prepare decisions and then make them in the formal way.

In this paper, we consider analysing the post-industrial CTPs from the viewpoint of a single DM as an element of its strategic planning. We analyse the problem from the viewpoint of the "Saturn" Museum managers.

4.2 Identifying Objectives

Based on the analysis of the nature of the problem, the decision context, an object and a preference direction, we can distinguish the following objectives:

1) Strategic objective: a new way of promoting, preserving and caring for the remaining post-industrial heritage in the Czeladź Commune.
2) Fundamental objectives: presenting a set of alternative decision variants of a product promoting post-industrial potential, enriching the local tourist offer, development of industrial and technical facilities tourism.
3) Means objectives: creating a new form of promotion of Czeladź Commune, finding other traces and remains of post-industrial heritage that have not been discovered so far.

The first objective was built deriving from the overall vision and mission of the new CTP on post-industrial heritage. The strategic concept of the new product assumes it allows disseminating information about the post-industrial heritage of two former coal mines in Czeladź. The fundamental objectives detailed the overarching strategic objective related to the vision and mission. Fundamental objectives are related to the instruments (tools, ways) based on alternative decision variants and potential solutions to the given problem. Means objectives reflect the intentions of development and improvement of strategic objective and are formulated on the basis of the analysis of success factors. The implementation of intermediate objectives leads to the main goal. In this case, the means objectives represent the possibilities of other, additional implementations and activities related to the general promotion of post-industrial CTP.

4.3 Establishing Criteria

There are multiple attributes and multiple criteria character of CTPs in general due to many typologies and classifications of CTPs in literature [1, 20, 27, 33, 37]. We analysed them and prepared a standardised set of 14 attributes/criteria, which will be very helpful to create (in the next step) a subset of key criteria to evaluate the CTP regarding local post-industrial heritage, such as:

- C1: Attractiveness of the product from the point of view of tourists.
- C2: Innovation in product development.
- C3: New technologies used in product development and its promotion.
- C4: Economic and social importance for the development of the region.
- C5: Relationships to events or traditions related to post-industrial time.
- C6: Authenticity (how well the product describes the post-industrial time).
- C7: Uniqueness (how original the product is).
- C8: Impact on general tourists infrastructure of the region.
- C9: Stimulation of tourist events in the region.
- C10: Stimulation of cultural events.
- C11: Providing new experiences, emotions, social contacts.
- C12: Enhancing promotion of the region, creating the region's image.
- C13: Providing educational impact for the users.
- C14: Shaping local/regional identity.

The DMs and stakeholders need to select out of the list above the key criteria that will best match the specificity and context of the CTPs creation process related to the promotion of post-industrial heritage. Properly designed surveys may help in establishing such a set of key criteria. Assuming that DMs and stakeholders evaluate the necessity of including each of these criteria in the set of key criteria using the 5-point Likert scale, their evaluations may be then aggregated (e.g. using simple Borda rule). This will result in a list of the most important criteria from the group viewpoint. Naturally, a cut-off point in selecting the final number of key criteria needs to be discussed with DMs and stakeholders.

In our case, the DM identified the subset of the most important criteria in brainstorming. They are shown in Table 1.

Table 1. List of key criteria evaluating CTP related to post-industrial heritage

Criterion code	Symbol and name of standardised criteria	Characteristics of the criteria
NEW_TECH	C3: New technologies in product development and its promotion	Does the product use new technologies, such as beacons, QR codes, mobile applications, web 2.0/3.0 technology, travel planners, geotagging, multimedia platforms, ICT systems, and e-books?
ECON	C4: Economic and social importance of the product for the development of the region	What is the product's economic and social importance for the development of the region? This includes (1) economic potential of the product; (2) product image, i.e. perception of the offer on the tourist market; (3) economic effect; (4) integration of the local community; (5) estimated future volume of tourist traffic; (6) tourist destination area
INFR	C8: General infrastructure	How much the product influences the development of infrastructure such as (1) tourist facilities; (2) recreational attractions; (3) accommodation base; (4) food and entertainment facilities; (6) transport and communication accessibility, (7) transport at the destination; (8) tourist and sports equipment and its rentals; (9) souvenirs shops
EVENTS	C10: Cultural events	Is the planned product conducive to such events as festivities, picnics, festivals, exhibitions, etc.
EXPER	C11: New experiences, emotions and impressions, new social contacts	Does the product have a positive effect on (1) getting to know the place, attraction, value, heritage; (2) excitement, fascination with the visited place; (3) establishing a relationship with people who experience and feel alike
PROM	C12: Promotion of the LGU	Is the product promoting the area of LGU, increasing the value of the LGU's tourist offer and building its image? Is it strengthening the competitiveness of the LGU on the regional market of tourist services?

(*continued*)

Table 1. (*continued*)

Criterion code	Symbol and name of standardised criteria	Characteristics of the criteria
EDU	C13: Education	Does the product affect the quality of the educational offer (e.g. giving the possibility of conducting thematic lessons or creating educational trials)

Source: own

4.4 Establishing the Set of Actions and Problematic

The variety of the categories, types and instances of CTPs (see Sect. 2.2) makes the possibility of building multi-products of cultural tourism. The presented problem is an ordering problem, we aim to present to DMs a list of possible decision alternatives with their evaluation, which allow them to get a deeper insight into the performance and quality of the possible solutions considered. It is an organisationally and technically challenging problem that deserves separate considerations, which lay behind the scope of this paper. Here, in Table 2, we present the characteristics of ten exemplary variants of CTP promoting the post-industrial heritage of the Czeladź Commune designed by our DM in brainstorming.

Table 2. Examples of alternative decision variants of CTP related to promotion of post-industrial cultural heritage occurring in the Czeladź Commune

Alternative	Combination in relation to the multidimensionality of the product	Description of the alternative variant of CTP
Route of Postindustria (ROUTE)	Product-route in real form and/or Product-thing in a hybrid form	Thematic cultural route leading through the most important points (places) of post-industrial heritage. The route also consists of dedicated, thematic sub-routes and educational trails concerning the technical monuments (machines, devices) and residential architecture

(*continued*)

Table 2. (*continued*)

Alternative	Combination in relation to the multidimensionality of the product	Description of the alternative variant of CTP
Postindustria Family Festivals (FESTIVALS)	Product-event in a real form and/or Product-event in a hybrid form	Thematic tourist and cultural events containing such attractions as educational workshops, outdoor family games, do it yourself (DIY) workshops, multimedia presentation of places, traces, artefacts
Postindustria Family Rally (RALLY)	Product-services set in real form	Thematic annual sports, tourist and culture event with elements of learning (workshops) about post-industrial culture
Postindustria Quest of Czeladź (QUEST)	Product-route in real form with questing and/or product-route in hybrid form with questing	Questing of post-industrial cultural heritage; outdoor game solving puzzles, tasks, quizzes, and finding the password
Postindustria Museum (MUSEUM)	Product-object in real form and/or product-object in hybrid form	Temporary, cyclical (once a year) exhibitions at the "Saturn" Museum and Contemporary Art Gallery "Elektrownia"
"Terra Postindustria" (AREA)	Product-area in the real form	Thematic geographically determined area of the former two coal mines, their patron estates and other infrastructure sites with routes, trails, questing games, cultural tourism facilities
"Postindustria Story" (STORY)	Product-service in real form and/or Product-service in hybrid form	Thematic story-based guided tour of the entire area related to the two mines and their heritage, divided into several thematic sections: (1) technical monuments, (2) residential architecture, (3) recreation, entertainment and (4) everyday life of mine workers, customs, rituals

(*continued*)

Table 2. (*continued*)

Alternative	Combination in relation to the multidimensionality of the product	Description of the alternative variant of CTP
Portfolio product A (PORTFOLIO A)	Material good (thing) + service + route	Map of post-industrial attractions, guide service of the most important attractions, thematic route through the most important post-industrial attractions
Portfolio product B (PORTFOLIO B)	Event + services set + virtual route	Thematic tourist and cultural festivities, picnics, festivals, exhibitions, etc. Thematic, sports, tourist rally with elements of learning about post-industrial culture. Virtual route on the web
Portfolio product C (PORTFOLIO C)	Product-thing in multimedia form + virtual service + virtual route	Interactive map of attractions (with photos, videos, graphics, animations), including virtual tour combined with the audiobooks through the virtual route on the website

Source: own

4.5 Identifying the State of Nature

For this problem, we consider that all criteria can be measured in a deterministic way, without the influence of non-controllable factors. Therefore, we do not consider the presence of stochastic variables with states of nature in our approach. In Sect. 5.1, we show how we could measure each of these criteria.

5 Analysing Preferences of the "Saturn" Museum Managers

5.1 Preference Modelling

In the preference modelling step, a significant question regarding the type of rationality that is most adequate to the DM should be raised. If the answer is—non-compensatory, a preliminary selection of non-compensatory methods (e.g. outranking methods) should be applied. If the answer is—compensatory, compensatory methods (such as MAUT (Multi-Attribute Utility Theory) or MAVT (Multi-Attribute Value Theory) should be in use (De Almeida et al., 2015). The question of selection of type and instance of the MCDM/A method is vital since choosing an inappropriate technique that would not meet the contextual requirements as well as the cognitive capabilities of DMs may result in false results, different for any two selected techniques (see, e.g. [21, 32]).

For our problem, compensatory rationality is assumed since the DM is willing to perform tradeoffs amongst criteria so that a better performance in another criterion can compensate for worse performance in another criterion. Therefore, in modelling preferences of the management of the "Saturn" Museum in our problem, we use the FITradeoff method [4, 19] and the DSS based on it – FU-TXMMO-WF1. This system elicits the scale constants of the additive model in a flexible and interactive way for the problems of choice and ranking. It allows the use of linear and non-linear value functions, conducting holistic evaluations to inform dominance relations between the alternatives of the problem, and performing sensitivity analysis of the results [11].

The method will be used using the subjective perception of the alternatives' performances regarding all evaluating criteria expressed by our DM in a consequence matrix. The management pre-evaluated and agreed upon the consequences in a free discussion using a 5-point Likert scale (in which 1 is the worst possible evaluation and 5 is the best possible evaluation). The consequences matrix obtained this way is shown in Table 3.

Table 3. The matrix of consequences for the post-industrial CTPs

Alternatives	KC1	KC2	KC3	KC4	KC5	KC6	KC7
A1	4	3	4	1	3	5	4
A2	3	3	3	5	4	3	3
A3	3	3	3	2	3	3	3
A4	4	3	3	2	4	4	4
A5	3	3	3	3	2	3	5
A6	3	4	4	2	3	4	4
A7	2	2	3	2	3	3	4
A8	3	3	4	2	3	4	4
A9	4	3	4	5	4	4	4
A10	5	3	4	2	4	3	3

Source: own

5.2 Intracriteria, Intercriteria and Alternatives Evaluation

In the FITradeoff method, steps 7, 8, and 9 of the framework are conducted jointly in the DSS. In order to support DM in their analyses, we applied the FU-TXMMO-WF1, which is based on the FITradeoff method. The criteria were ranked first using the option of ranking the criteria scaling constants in DSS. Then the series of questions was asked to compare hypothetical pairwisely (see Fig. 1) to produce the ranges for criteria weights (Fig. 2) and the rank order of alternatives.

In our case, the most important criterion is "Promotion of the LGU (PROM)" and the least important one is General infrastructure (INFR). During the elicitation phase, after asking eight questions the system was not able to determine the complete order of alternatives, as two of them, i.e. alternative "Route of Postindustria (ROUTE)" and "Portfolio product B (PORTFOLIO B)" were still considered incomparable. Luckily, the ninth question allowed to produce the complete order of alternatives identifying the following rank order of alternatives: ROUTE → PORTOFOLIO B → QUEST → AREA and PORTFOLIO C → PORTFOLIO A → FESTIVALS → MUSEUM → RALLY → STORY.

This ranking shows us that the best option turns out to be the "Route of Postindustria (ROUTE)" option concerning the creation of the thematic cultural route leading through the most important points (places) of post-industrial heritage. Another good option is "Portfolio product B (PORTFOLIO B", a portfolio product that consists of: (1) thematic tourist and cultural festivities, picnics, festivals, exhibitions, etc.; (2) thematic, sports, tourist rally with elements of learning about post-industrial culture; (3) virtual route on the website. The worst alternative is "Postindustria Story" which is a thematic story-based guided tour of the entire area related to the two mines and their heritage.

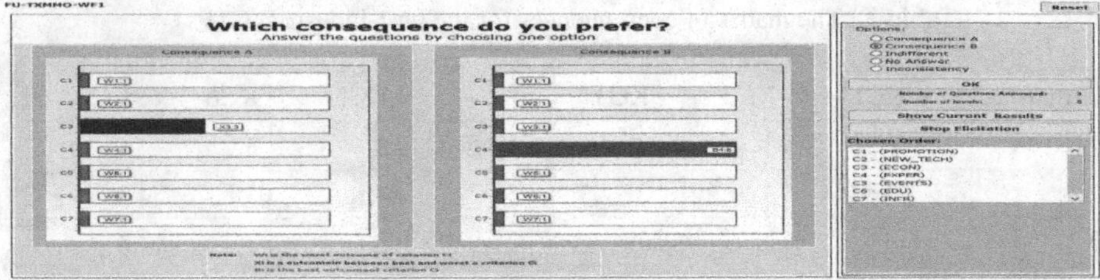

Fig. 1. Elicitation of preference for hypothetical alternatives Source: FU-TXMMO-WF1 – FITradeoff method based system

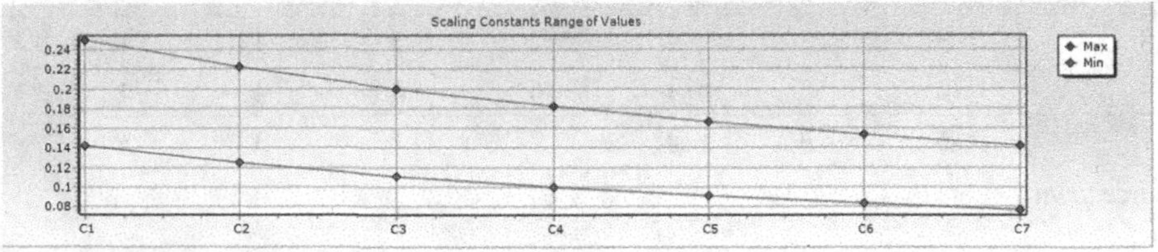

Fig. 2. Ranking the criteria Source: FU-TXMMO-WF1 – FITradeoff method based system

6 Finalisation

6.1 Conducting a Sensitivity Analysis

The results should be checked for their robustness, assuming changes in the parameters of the model and its input data. This step may indicate that the recommendation is either: robust or sensitive to the input data or the model features. It may also show that the results from step 9 should be reevaluated due to false assumptions or input data or inadequate simplification in the model [3]. The FITradeoff method based DSS has the functionality for a sensitivity analysis. Suppose we conduct sensitivity analysis for alternatives' consequences in criterion "Promotion of the LGU (PROM)" – assuming the possible ±20% of changes in values (changes of 1 level in 5-point Likert scale evaluation). The results of such sensitivity analysis are shown in Fig. 3.

Percentage of times that the alternative was ordered in the position:

Alternative/Position	1	2	3	4	5	6	7	8	9	10
(ROUTE)	57.40%	17.00%	20.70%	4.80%	0.10%	0.00%	0.00%	0.00%	0.00%	0.00%
(FESTIVALS)	0.00%	0.20%	1.60%	10.20%	8.50%	18.70%	15.70%	32.00%	12.30%	0.80%
(RALLY)	0.00%	0.10%	0.50%	0.80%	11.70%	22.00%	41.70%	19.80%	3.20%	0.20%
(QUEST)	14.30%	20.50%	45.20%	9.90%	8.10%	1.90%	0.10%	0.00%	0.00%	0.00%
(MUSEUM)	0.00%	0.10%	1.00%	9.20%	18.50%	42.50%	20.30%	3.50%	4.60%	0.30%
(AREA)	5.40%	11.70%	17.40%	43.20%	12.10%	9.60%	0.60%	0.00%	0.00%	0.00%
(STORY)	0.00%	0.00%	0.00%	0.10%	1.30%	10.30%	24.80%	44.50%	16.80%	2.20%
(PORTFOLIO A)	4.70%	10.00%	9.70%	14.30%	48.80%	11.20%	1.30%	0.00%	0.00%	0.00%
(PORTFOLIO B)	25.50%	50.40%	15.40%	6.00%	2.50%	0.20%	0.00%	0.00%	0.00%	0.00%
(PORTFOLIO C)	1.20%	6.60%	21.00%	58.20%	5.20%	1.60%	1.80%	2.70%	1.70%	0.00%

Fig. 3. Sensitivity analysis for issue of promotion of the LGU (PROM) Source: FU-TXMMO-WF1 – FITradeoff method based system

One can see that the alternative "Portfolio product B (PORTFOLIO B)" (with the probability of 25,50%) and "Postindustria Quest of Czeladź (QUEST)" (with probability of 14,30%) could be at first place in ranking of alternatives. This is an important information, especially if the original performances of alternatives were the subject of the compromise among the managers representing our DM and could be changed inn further discussion. We can see that this could affect the final ranking of alternatives.

Similar sensitivity analysis can be performed for other criteria providing our DM with additional information on the stability of the ranking obtained.

6.2 Drawing up Recommendations and Implementation Actions

The finalisation is conducted by analysing the final results and producing the report for the DMs and stakeholders, with the final recommendations. Note, however, that in this paper, we analysed the problem of analysing the possible CTPs from the viewpoint of the single DM, as an element of their preparation to broader discussion with other DMs and stakeholders. With the information about the potential importance of the evaluation criteria (their ranges) and the resulting rank order of alternatives, the "Saturn" Museum managers are ready to start their prenegotiation preparation for the forthcoming negotiations with other stakeholders. Knowing the best solutions for themselves, the managers may now efficiently plan their negotiation and concession strategy, as recommended by the theory of negotiation analysis [36].

7 Conclusions

In this paper, we identified the post-industrial heritage of Czeladź Commune and provided a decision analysis of the performance of potential CTPs that may promote it. The FITradeoff method and corresponding electronic DSS were used to perform the decision analysis. It turned out that finding a solution using FITradeoff quite fast and allow easily to take into account the subjectivism of our DM in defining their preferences. It should be highlighted that the FITradeoff method operates based on partial information provided by the DM, which allows the elicitation process to be carried out with less cognitive effort, compared to traditional complete information methods, such as the classic tradeoff. The visualisation of possible tradeoffs was considered by them as easy and intuitive and allow to build the ranking of alternatives in nine easy steps. It confirms earlier simulation studies [31] that the convergence of the method is fast.

The three alternatives: "Route of Postindustria", "Portfolio product B", and "Postindustria Quest of Czeladź" can be seen as a set of very good solutions. The similar features and characteristics of these alternatives make them have a very similar influence on the key important criteria for DM. The common idea of these variants is the promotion of post-industrial heritage by: (1) active cultural tourism; (2) providing routes or paths created ad hoc (as quest); (3) learning about heritage objects. These are the alternatives most preferable for the "Saturn" Museum and should be set as the aspiration levels in the forthcoming negotiations with other future stakeholders in the problem of creating post-industrial CTP in Czeladź. Additionally, by having the complete ranking of offers, the other CTPs were identified as the potential concessions that may be used in the negotiations.

The problem, however, requires further analyses to provide a comprehensive solution for all possibly involved stakeholders. In particular, we should focus on designing the process of construction of different variants of CTPs by using some formal procedures, deriving from portfolio building theory and using knapsack models. Further, the mechanism for supporting all stakeholders should be developed deriving from group decision-making and taking into account individual stakeholders' cognitive abilities and information processing styles. This concept could involve, for example, an a-few-step approach. In the first step, the stakeholders could take properly prepared psychological tests to determine their dominant information processing style. Next, the characteristic features of a given information processing style would be compared with the specificity of procedures of MCDA methods to find the best possible selection of MCDA methods to be applied to support a specific subgroup of respondents. The use of this selected method by individual respondents belonging to a given subgroup will be one of the steps in the further stages (in the second phase) of the study.

References

1. Abdurahman, A.Z.A., Ali, J.K., Khedif, L.Y.B., Bohari, Z., Ahmad, J.A., Kibat, S.A.: Ecotourism product attributes and tourist attractions: UiTM undergraduate studies. Procedia Soc. Behav. Sci. **224**, 360–367 (2016)
2. Alfieri, A., et al.: A multi-objective tabu search algorithm for product portfolio selection: a case study in the automotive industry. Comput. Ind. Eng. **142**, 106382 (2020). https://doi.org/10.1016/j.cie.2020.106382
3. De Almeida, A.T., Cavalcante, C.A.V., Alencar, M.H., Ferreira, R.J.P., De Almeida-Filho, A.T., Garcez, T.V.: Multicriteria and Multiobjective Models for Risk, Reliability and Maintenance Decision Analysis. International Series in Operations Research & Management Science, vol. 231. Springer, New York (2015). https://doi.org/10.1007/978-3-319-17969-8

4. De Almeida, A.T., de Almeida, J.A., Costa, A.P.C.S., de Almeida-Filho, A.T.: A new method for elicitation of criteria weights in additive models: flexible and interactive tradeoff. Eur. J. Oper. Res. **250**(1), 179–191 (2016)
5. de Almeida, A.T., Frej, E.A., Roselli, L.R.P.: Combining holistic and decomposition paradigms in preference modeling with the flexibility of FITradeoff. CEJOR **29**(1), 7–47 (2021). https://doi.org/10.1007/s10100-020-00728-z
6. Archival collection of the "Saturn" Museum in Czeladź. Project of the mine building. Inventory no. MS/HG/153. Accessed 31 July 2021
7. Archival collection of the "Saturn" Museum in Czeladź. Album of the Mining and Industrial Society "Saturn". Inventory no. MS/HG/461. Accessed 31 July 2021
8. Ban, O.: The opportunity of indirect determination of the importance of the attributes of the tourist product in evaluating the consumer's satisfaction. In: 2012 International Conference on Economics, Business and Marketing Management. IPEDR, vol. 29. IACSIT Press, Singapore (2012)
9. Binek-Zajda, A., Lazar, S., Szaleniec, I.: Coal mine and workers' settlement "Saturn". History, architecture, people. Public Society "Czeladź" and the "Saturn" Museum in Czeladź, Czeladź (2016). (in Polish)
10. Blázquez, J., Molina, A., Esteban, Á.: Key quality attributes according to the tourist product. Eur. J. Tour. Res. **5**(2), 166–170 (2012)
11. Borba, L.B., Leal, M.J.V., Ribeiro, M.L.S., Santo, P.P.P.: Flexible and interactive tradeoff elicitation for additive model with sensitivity analysis (FU_TXMMO_WF1) – intraweb version. Practical User Guide. Universidade Federal de Pernambuco, Center for Decision Systems and Information Development, Recife (2021)
12. Bujdosó, Z., et al.: Basis of heritagization and cultural tourism development. Procedia Soc. Behav. Sci. **188**, 307–315 (2015)
13. Burkart, A.J., Medlik, S.: Tourism: Past, Present and Future. Heinemann, London (1981)
14. Chmielewska, M., Lamparska, M., Pytel, S., Jurek, K.: Patronage housing estates in Zagłębie. Tourist route project. Association for the Protection of Natural and Cultural Heritage "MOJE MIASTO", Będzin (2016). (in Polish)
15. Chou, T.-Y., Hsu, C.-L., Chen, M.-C.: A fuzzy multi-criteria decision model for international tourist hotels location selection. Int. J. Hosp. Manag. **27**(2), 293–301 (2008)
16. Davies, B.: The role of quantitative and qualitative research in industrial studies in tourism. Int. J. Tour. Res. **5**(2), 97–111 (2003)
17. Domaszewski, K.: From a trip to Saturn. In: Zeszyty Czeladzkie 7, Czeladź (2000). (in Polish)
18. Eden, C.: Analysing cognitive maps to help structure issues or problems. Eur. J. Oper. Res. **159**(3), 673–686 (2004)
19. Frej, E.A., de Almeida, A.T., Costa, A.P.C.S.: Using data visualization for ranking alternatives with partial information and interactive tradeoff elicitation. Oper. Res. **19**(4), 909–931 (2019). https://doi.org/10.1007/s12351-018-00444-2
20. Fuadillah, N., Murwatiningsih, M.: The effect of place branding, promotion and tourism product attribute to decision to visit through the destination image. Manag. Anal. J. **7**(3), 328–339 (2018)
21. Guitouni, A., Martel, J.-M.: Tentative guidelines to help choosing an appropriate MCDA method. Eur. J. Oper. Res. **109**(2), 501–521 (1998)
22. Hammond, J.S., et al.: Smart Choices: A Practical Guide to Making Better Decisions. Broadway Books, New York (2002)
23. Kasimoğlu, M., Aydin, H.: Strategies for Tourism Industry – Micro and Macro Perspectives. InTech, Rijeka (2012)
24. Keane, M.J.: Quality and pricing in tourism destinations. Ann. Tour. Res. **24**(1), 117–130 (1997)

25. Kurek, R.: The beginnings and development of industry in Czeladź. In: Drabina, J. (ed) History of Czeladź 1, Czeladź (2012). (in Polish)
26. Lazar, S., Binek-Zajda, A.: Piaski housing estate. History and architecture. Public Society "Czeladź" and the "Saturn" Museum in Czeladź, Czeladź (2015). (in Polish)
27. Logunova, N., Kalinkina, S., Lazitskaya, N., Tregulova, I.: Methods and criteria for assessing the effectiveness of cruise tourism development. In: VIII International Scientific Conference Transport of Siberia. IOP Conference Series: Materials Science and Engineering, vol. 918 (2020)
28. Lohmann, M.: New demand factors in tourism. Paper to be Presented to the European Tourism Forum, Budapest, 15 October 2004 (2004)
29. Mason, P.: Tourism Impacts. Planning and Management. Routledge, New York (2016)
30. Medlik, S., Middleton, V.T.C.: Product formulation in tourism. Tour. Mark. **13**, 138–154 (1973)
31. Mendes, J.A.J., Frej, E.A., de Almeida, A.T., de Almeida, J.A.: Evaluation of flexible and interactive tradeoff method based on numerical simulation experiments. Pesquisa Operacional **40**, 1–25 (2020)
32. Moshkovich, H.M., et al.: Influence of models and scales on the ranking of multiattribute alternatives. Pesquisa Operacional **32**, 523–542 (2012)
33. Nair, M.B., Ramachandran, S., Shuib, A., Syamsul, H.M.A., Nair, V.: Multi-criteria decision making approach for responsible tourism management. Malays. For. **72**(2), 135–146 (2012)
34. Panasiuk, A.: From basic tourism products to a comprehensive offer of a tourism area. Barometr Regionalny **15**(1), 17–24 (2017)
35. Puchinger, J., et al.: The multidimensional knapsack problem: structure and algorithms. INFORMS J. Comput. **22**(2), 250–265 (2010). https://doi.org/10.1287/ijoc.1090.0344
36. Raiffa, H., et al.: Negotiation Analysis: The Science and Art of Collaborative Decision Making. Harvard University Press (2002)
37. Ramírez-Guerrero, G., García-Onetti, J., Chica-Ruiz, J.A., Arcila-Garrido, M.: Concrete as heritage: the social perception from heritage criteria perspective. The Eduardo Torroja's work. Int. J. Des. Nat. Ecodyn. **15**(6), 785–791 (2020)
38. Russo, A.P., van der Borg, J.: Planning considerations for cultural tourism: a case study of four European cities. Tour. Manag. **23**, 631–637 (2002)
39. Smith, S.L.J.: The tourism product. Ann. Tour. Res. **21**(3), 582–595 (1994)
40. State Archive in Katowice – The Mining and Industry Society "Saturn" (1862–1900), (1900–1944), (1945–1953). Accessed 31 July 2021
41. State Archive in Katowice – The Nameless Society of the Coal Mines "Czeladź" in Czeladź-Piaski (1919–1922). Accessed 31 July 2021
42. Stefano, N.M., Casarotto Filho, N., Barichello, R., Sohn, A.P.: Hybrid fuzzy methodology for the evaluation of criteria and sub-criteria of product-service system (PSS). Procedia CIRP **30**, 439–444 (2015)
43. Stokes, R.: Tourism strategy making: insights to the events tourism domain. Tour. Manag. **29**(2), 252–262 (2008)
44. Szromek, A.R., Herman, K.: A business creation in post-industrial tourism objects: case of the industrial monuments route. Sustainability **11**(5), 1–17 (2019)
45. Vucetic, A.: Impact of tourism policy on development of selective tourism (2009). https://ssrn.com/abstract=3579681. Accessed 31 July 2021
46. Weber, F., Taufer, B.: Assessing the sustainability of tourism products – as simple as it gets. Int. J. Sustain. Dev. Plan. **11**(3), 325–333 (2016)
47. Williams, P.W., Penrose, R.W., Hawkes, S.: Shared decision-making in tourism land use planning. Ann. Tour. Res. **25**(4), 860–889 (1998)

48. Wong, F.K.K., Fung, T.: Ecotourism planning in Lantau Island using multiple criteria decision analysis with geographic information system. Environ. Plann. B Urban Anal. City Sci. **43**(4), 640–662 (2016)
49. Yu, X., Xu, H.: Cultural heritage elements in tourism: a tier structure from a tripartite analytical framework. J. Destin. Mark. Manag. **13**, 39–50 (2019)

The Worth of Cardinal Information in MCDM – a Guide to Selecting Weight-Generating Functions

Mats Danielson[1,2] and Love Ekenberg[2,1](✉)

[1] Department of Computer and Systems Sciences, Stockholm University, PO box 7003, SE-164 07 Kista, Sweden
mats.danielson@su.se, ekenberg@iiasa.ac.at
[2] International Institute for Applied Systems Analysis IIASA, Schlossplatz.1, AT-2361 Laxenburg, Austria

Abstract. There exist a variety of methods for extracting weights and values in multi-criteria decisions based on different rankings. However, it is difficult to determine which the useful ones are and how they correspond to the decision makers' perceptions, if at all. How do we know that what we are using is really significant, especially in situations when the decision bases are vague? One category of methods that has proved relatively successful is to use so-called generated surrogate weights that are, in some sense, meant to represent rankings and there are various suggestions as to how best to distil them from input information. In this paper, using a number of simulations, we review some leading algorithms for automatic weight generation without external parameters besides cardinal and ordinal rankings and provide some guidelines for selecting a surrogate weight-generating function for MCDM applications, in ordinal as well as cardinal information settings. We also propose an alternative with some attractive properties compared with the existing ones.

Keywords: Multi-criteria decision analysis · Surrogate numbers · Robustness · Rank order

1 Introduction

A common underlying measurement mechanism in multi-criteria decision analysis (MCDM) is Multi-Attribute (Value or) Utility Theory (MAVT / MAUT), where a common model is the additive evaluation one, i.e., $V(a) = \sum_{i=1}^{m} w_i v_i(a)$, where $V(a)$ is the overall value of alternative a, $v_i(a)$ is the value of the alternative under criterion i, and w_i is the weight of this criterion. This model is fraught with some difficulties as many emphatically has pointed out [1], not least because it requires information that the decision-maker does not have access to, which can result in decisions being based on estimates or pure conjecture. This has been dealt with in various ways by allowing imprecise information, e.g. in the form of intervals, distributions, or ordinal or cardinal

A. T. de Almeida and D. C. Morais (Eds.): INSID 2021, LNBIP 435, pp. 20–35, 2021.
https://doi.org/10.1007/978-3-030-91768-5_2

rankings of criteria and alternatives. This article focuses on the latter approach, more precisely to the issue of elicit preference information.

To represent ordinal and cardinal relations, one important class of methods uses auto-generated weights and (sometimes) alternative values to express some kind of plausible interpretation of the decision-makers' preference orders. Needless to say, it is not immediately obvious which of the various proposals to choose or what their characteristics are relative to other proposals. Barron and Barrett [2] made an important contribution in systematising the evaluation of some suggestions by utilising simulations. The idea is to construct both automatic random weights and a kind of "true" reference weights based on underlying distributions and then study how well the results correlated. However, the method in [2] has a weakness in that the results are highly dependent on the distribution used to create the weight vectors. There are also other issues involved, such that the employment of ratio weight procedures can be difficult due to response errors [3]. In any case, one category of surrogate number generation is to derive them from ordinal importance information [4, 5], utilising rank orders, i.e., ordinal information, whereafter these orderings are transformed into numerical weights in correspondence with the information. Such methods include rank sum (RS) weights and rank reciprocal (RR) weights [6], as well as centroid (ROC) weights [7]. The information loss, when using ordinal information only, could be problematic why cardinal orderings have been suggested as an alternative for utilising further information that might be present in decision situations.

In this article, we review and measure some leading well-known automatic weight-generating functions that do not contain external parameters, i.e. they do not require any other information from a decision-maker than the ranking. We thus discuss the properties of a number of surrogate number methods from a robustness perspective as well as an efficiency viewpoint. Using a simulation approach, we compare a set of ranking methods for weights and their relevance. In Sect. 2, we give a general overview of some techniques for representing ranking methods. Section 3 covers cardinal rankings. In Sect. 4, we describe the simulation approach and provide an overview of six cardinal methods and their robustness properties. Section 5 concludes the paper.

2 Automatic Weight-Generating Functions

A number of different preference elicitation methods have been proposed over the years, such as scoring points (point allocation, PA) and direct rating (DR) methods. In PA, the decision-maker provides a score that is distributed across or divided between the criteria. After the weights for N–1 criteria are distributed, the N criterion is determined, i.e. there are N–1 degrees of freedom (DoF). In contrast, DR methods do not put a limit on these scores and each score is simply divided by the total, so there are N degrees of freedom for N criteria.[1] Regardless of which method is used, some form of weight distribution is thus implied. So-called rank sum (RS) and rank reciprocal (RR) weights are discussed in

[1] In mathematics and in simulations, as well as in human real-life reasoning, the resulting distribution of weights differs if a) the weights are first assigned disregarding the requirement to sum to one and then subsequently normalised by dividing by their sum; or b) all except one weight are assigned and the remaining one receives what is left to make them sum to one.

Stillwell et al. [6] and suggested as alternatives to ratio-based weight schemes. RS means that the ranking should be reflected directly in the weights. Given N criteria weights ($i = 1,\ldots,N$) and the constraints $w_1 > w_2 > \ldots > w_i > \ldots > w_N \geq 0, \sum w_i = 1$, the RS surrogate weights are given by:

$$w_i^{RS} = \frac{N+1-i}{\sum_{j=1}^{N}(N+1-j)} \tag{1}$$

RR is instead based on the reciprocals of the rank order:

$$w_i^{RR} = \frac{1/i}{\sum_{j=1}^{N}\frac{1}{j}} \tag{2}$$

Barron [7] suggested the ROC (rank order centroid) weights as the average of the corners in the simplex defined by the constraints, i.e. the weights are the components of the centroid of the simplex:

$$w_i^{ROC} = 1/N \sum_{j=i}^{N}\frac{1}{j} \tag{3}$$

The properties of RS, RR, and ROC are discussed in [8], and it can be seen that their performance is strongly dependent on quite strong assumptions regarding the decision-makers. RS performs best when we assume N DoF and RR as well as ROC under the assumption of N–1 DoF. Danielson and Ekenberg [8] therefore suggested a weight generation combining the properties of RS and RR while considering different degrees of DoF:

$$w_i^{SR} = \frac{1/i + \frac{N+1-i}{N}}{\sum_{j=1}^{N}\left(1/j + \frac{N+1-j}{N}\right)} \tag{4}$$

There is also a range of cardinal-based alternatives as discussed e.g. in [9]. For instance, Simos [10, 11] proposed a quite popular visual method for easily representing criteria rankings, possibly extended with some cardinality. A decision-maker group is given a set of coloured cards with the criteria names written on them as well as a set of blank cards. The coloured cards are placed in preference order and the white cards are placed in between the coloured cards to provide information on preference strengths, where a constant value difference, u, between two consecutive cards is assumed. For instance, a white card between two consecutive coloured ones denotes a difference of $2 \cdot u$. From a set of ordered cards, the normalised weights can then be determined. The Simos method is called SI in the study below and implemented according to [10, 11].

Regardless of the method chosen, its quality will depend on assumptions of the mental model of the decision-maker's thought process. To investigate this further, we divide the automatic weight generation methods into three categories depending on which model of the decision-maker's thought process they adhere to. In the first category, which we will call category **N1**, resides the methods that bet on the decision-makers using an N–1 DoF way of reasoning and thus aligns their weighting functions to work the most properly in that case. The **N1** category consists, in this article, of ROC and RR. As we will see in the results section, they perform markedly better in an *N*–1 scenario and worse in an *N* scenario. The second category, which we will call category **N**, contains the methods that instead bet on decision-makers using an *N* DoF way of reasoning. The RS and Simos methods fall into that category in this article. As opposed to the **N1** category, these methods perform better in an *N* scenario than in an *N*–1 one. Finally, we have methods that are designed to work in both scenarios and in any mixture thereof. We will call this category the **M** category (M for mixed), and besides the SR weight function introduced in [8], we will in this paper introduce another function with the same rationale but with somewhat differing properties. The key idea is still to obtain a high level of robustness, but this time not only with respect to differing degrees of freedom as previously, but also with respect to the loss of information when cardinality is decreased (or, conversely, the gain of information when it is increased). In a way similar to the construction of SR, the *SUMROC* weight function (sometimes denoted SC in this article for short) is composed of a combination of the RS and ROC functions (the SR function was an amalgamation of RS and RR), retaining the idea of combining an *N*–1 aligned function with an *N* one. In this setting, the ROC function has the role of an *N*–1 DoF catering function and RS has the role of an *N* DoF catering one. The combination of them is therefore a candidate for a robust automatic weight-generating function that caters to the whole spectrum of decision processes with *N* and *N*–1 DoF as its endpoints, much in the same way as SR has been proven to be, but its efficacy remains to be shown. Section 4 will address this question. SC is defined as:

$$w_i^{\mathrm{SC}} = \frac{\left(\sum_{j=i}^{N} \frac{1}{j}\right) + N + 1 - i}{N + \sum_{j=1}^{N} j} \tag{5}$$

where the denominator is, as usual, the normaliser ensuring that the generated weights sum to one.

During interviews and focus groups with 139 decision-makers over two years, where each one of them was instructed to make an important decision in their lives and with each decision taking on average three weeks, we came to the conclusion that decision-makers vary vastly in the way they reason about criteria weights; some using thinking more akin to an *N* DoF model and others rather being closer to an *N*–1 model, while many did something in between since the two models could be seen as endpoints on a rather large continuum. While the studies reported in [12] and [13] did not have the issue of decision-maker DoF as their main focus (which instead was comparing the SMART, AHP, and CAR families of MCDM approaches), one big takeaway was the modes of reasoning each decision-maker displayed during the study. While some clearly described either having some kind of "putty" at their disposal to distribute (an *N*–1 model), others rather acknowledged that they awarded points to the criteria while worrying about if it was too much or too little at a much later stage (an *N* model). But the most common situation was for the decision-makers to describe something in between, being aware of a total weight sum (100%) that must be satisfied and keeping it in the back of their heads but not acting on or enforcing it until somewhat later in their thinking process.

3 The Weight of Cardinal Strength

Ordinal orders are of course relatively weak and there may often be more information in a decision situation. For example, decision-makers may have more or less strong preferences between the constituent criteria. The surrogate weights generated from pure ordinal rankings may therefore be unnecessarily misleading in that they do not make use of the information at hand. Therefore, there may be a case (like e.g. Simos) for extending ordinal rank orders to include strength relations. However, cardinal scales are usually more complicated for decision-makers to use than ordinary ones. For example, the ratio scales in the original AHP method (using a number of ratios) or the scoring in the SMART family (using multiple integers) require a precision that can be difficult to accomplish. There is therefore a compelling case for looking at some alternatives.

Assume that there exists an ordinal ranking of N criteria. In order to make this order into a stronger ranking, information should be given about how much more or less important the criteria are compared to each other. Such rankings also take care of the problem with ordinal methods of handling criteria that are found to be equally important, i.e. resisting pure ordinal ranking. In this paper, we will use the following notations for the strength of the rankings between criteria as well as some suggestions for a verbal interpretation of these:

$>_0$ Equally important.
$>_1$ Slightly more important.
$>_2$ More important (clearly more important).
$>_3$ Much more important.

In analogy with the ordinal weight functions in the previous section, counterparts using the concept of preference strength can straightforwardly be derived.

1. Assign an ordinal number to each importance scale position, starting with the most important position as number 1.
2. Let the total number of importance scale positions be Q (in a Simos terminology, the number of blanks and coloured cards). Each criterion i has the position $p(i) \in \{1,\ldots,Q\}$ on this importance scale, such that for every two adjacent criteria c_i and c_{i+1}, whenever $c_i >_{s_i} c_{i+1}$, $s_i = |p(i+1) - p(i)|$. The position $p(i)$ then denotes the importance as stated by the decision-maker. Thus, Q is equal to $\sum s_i + 1$, where $i = 1,\ldots,N-1$ for N criteria.

Then the cardinal counterparts to the ordinal ranking methods can be found by using this extension.

4 Investigation of Cardinal Strength for Automatic Weights

How should methods for cardinal preference ordering be validated? One way is to use simulations in analogy with previous studies on ordinal orders such as in [1, 14–16], which has become a kind of validation standard in the field. Here it is assumed that the decision-maker (unconsciously) has a set of "true" weights, which are unavailable in terms of exact numbers, but that they nevertheless exist in some abstract sense and against which the simulation results can be matched. The weakness, besides perhaps a questionable metaphysical assumption, is that the validation becomes heavily dependent on the random generation of vectors. For example, if we assume an N–1 DoF model, the components should sum to 1, while if we assume an N DoF model, all values in the vectors lie between 0 and 1 that are then subsequently normalised. One can, of course, as in [17] assume mixtures of this or assume completely different distributions, but the important observation here is that the validation itself depends on assumptions as strong as the assumptions regarding the cognitive performance of the decision subjects.

This, of course, raises the issue of the reliability of the validations. An N–1 DoF model assumes a homogeneous N-variate Dirichlet distribution, cf. [18]. If one starts from an N DoF model, one has a uniform distribution with N degrees of freedom that is normalised only in the next step, see e.g. [19] for details. We call the corresponding generator types N–1 generators and N generators, respectively. The background assumption about the cognitive notion of the decision-maker already has a large impact here. In [8] we discuss how ROC weights and N–1 generators on one hand and RS weights and N generators on the other have a considerable impact on the validation results. In general, we can rarely know whether particular decision-makers have mental N–1 or N DoF preference representations or anything in between these. Even less so in the case of a group of decision-makers who are trying to find a common model, why there must be a great deal of flexibility in this respect. In ranking mechanisms, the generation of surrogate weights must thus be able to handle both types of representation, as well as mixtures of them.

4.1 Generation Procedure

In [17], the following procedure is described for the assessment of different automatic weight generation functions. The same procedure has been used in this article.

1. For an N-dimensional problem, generate a random weight vector $\boldsymbol{t}$ with N components. This is called the true weight vector. Determine the order between the weights in the vector $\boldsymbol{t}$. For each method X', use the order to generate a weight vector $w^{x'}$.
2. Given M alternatives, generate $M \times N$ random values with value v_{ij} belonging to alternative j under criterion i.
3. Let w_i^x be the weight from weighting method **X** for criterion i (where **X** is either **X'** or $\boldsymbol{t}$). For each method **X**, calculate $V_i^x = \sum w_i^x v_{ij}$. Each method produces a preferred alternative A_x, i.e. the one with the highest V_i^x.
4. For each method **X'**, assess whether **X'** yielded the same decision (i.e. the same preferred alternative $A_{\mathbf{X}}$) as $\boldsymbol{t}$. If so, record a hit.

The hit rate is the number of times a weighting method got the same result as using the TRUE vector. Other possible measures of effectiveness are average loss of value and average percentage of the maximum value range, both of which however correlate strongly with the hit rate and therefore do not provide much more information regarding the validity of the respective methods.

4.2 Comparing Weight Methods

Similar to earlier studies, the comparative simulations were carried out with a varying number of criteria and alternatives. There were five numbers of criteria $N = \{3, 6, 9, 12, 15\}$ and five numbers of alternatives $M = \{3, 6, 9, 12, 15\}$ creating a total of 25 simulation scenarios of which we show nine in the tables below. Each scenario was run 10 times, each time with 10,000 trials, yielding a total of 2,500,000 decision situations generated. An N-variate joint Dirichlet distribution was employed to generate the random weight vectors for the N–1 DoF simulations and a standard round-robin normalised random weight generator for the N DoF simulations. Similar to [2], unscaled value vectors were generated uniformly.

All numbers in the tables below are given in ‰ (per mille) or ppt (parts per thousand), where 1000 would indicate a full occurrence frequency of 1 (or 100%). The analysis of the results starts with the ordinal case. In this case, all criteria are ranked in a strict order. For the ordinal case, as for all subsequent cases, the simulations were performed for N DoF scenario decision-makers, for N–1 ones, and for an equal mixture of the two.

The latter mixture is shown in the result tables below while the former are found for completeness reasons in Appendix B. It is argued above that since we do not know the decision method of a specific decision-maker, a good weight-generating function must be able to function in both scenario endpoints as well as anywhere in between. The best indicator for the analysis is the mixed DoF tables below. In Table 1, the results for the ordinal case are shown (only $>_1$ is allowed in our terminology). Note that for SI (Simos), the results are exactly the same as for RS. This is due to the fact that when SI is used on ordinal data, the algorithm coincides with that of RS. In the left column of Table 1, the category that a particular algorithm belongs to according to Sect. 2 is shown in parenthesis.[2]

Table 1. The ordinal frequency for the methods using a combination of DoFs

	3/3	3/15	6/6	6/12	9/9	12/6	12/12	15/3	15/15
ROC(N1)	888	783	823	789	798	818	785	868	772
RS(N)	888	787	835	802	817	836	804	884	795
RR(N1)	889	777	804	765	746	742	702	778	659
SR(M)	893	787	837	807	821	841	809	889	800
SC(M)	893	793	846	809	830	848	816	891	805
SI(N)	888	787	835	802	817	836	804	884	795

[2] In this and the following tables, the headings contain the notation A / B, denoting a decision situation having A criteria and B alternatives from the set {3, 6, 9, 12, 15}. The numbers in the set are selected to cover a broad range of decision situations encountered in real life. The tables show tendencies when the number of criteria and alternatives are being changed, but changing in steps smaller than 3 would not considerably impact the trends seen. This way, the tables are kept at reasonable size.

Next, the corresponding tables are shown for the semi-ordinal case (Table 2), where two criteria are allowed to be deemed equal, but with no blanks in between criteria ($>_0$ and $>_1$ are allowed in our terminology).

Table 2. The semi-ordinal frequency for the methods using a combination of DoFs

	3/3	3/15	6/6	6/12	9/9	12/6	12/12	15/3	15/15
ROC	903	805	844	813	817	832	802	878	787
RS	898	796	849	816	834	854	825	898	819
RR	895	781	816	775	763	761	721	802	679
SR	898	793	851	816	833	854	824	899	814
SC	905	807	863	830	849	868	841	908	833
SI	898	799	842	810	821	840	808	887	798

Following this, the next table (Table 3) shows the results when up to two blank positions (blank cards in ELECTRE/Simos terminology) are allowed. This entails that ranking symbols $>_0$, $>_1$, $>_2$, and $>_3$ are allowed. For completeness, the intermediary case of one blank (symbols $>_0$, $>_1$, and $>_2$) is shown in Appendix A, but these intermediate results do not change the conclusions in any respect.

Table 3. The cardinal frequency for the methods using a combination of DoFs

	3/3	3/15	6/6	6/12	9/9	12/6	12/12	15/3	15/15
ROC	891	772	813	776	804	788	752	837	738
RS	924	838	905	882	895	908	889	933	883
RR	896	785	755	707	666	656	599	713	552
SR	916	823	867	837	845	858	832	893	824
SC	920	831	909	885	905	914	894	939	891
SI	921	836	894	868	877	890	867	918	855

From Tables 1, 2 and 3, the gain in decision power from introducing cardinality can be inferred. Tables 4 and 5 show the difference in frequency between the cardinal cases and the ordinal case which serves as a base case in the comparison. Note that some of the gains are *negative*, i.e. there is a loss of decision power from introducing more information.

Finally, the last step before the analysis is to show the robustness of each weight-generating function. As argued in [9] and [12], since there is no way of knowing the DoF reasoning of a particular decision-maker, it is important for a viable weight method to cover the full spectrum from *N*–1 DoF to *N*. This robustness is measured by the difference in frequency between the two endpoint DoFs, and Tables 6, 7 and 8 show

Table 4. The gain in frequency from using semi-cardinal information

	3/3	3/15	6/6	6/12	9/9	12/6	12/12	15/3	15/15
ROC	15	22	21	24	19	14	17	10	15
RS	10	9	14	14	17	18	21	14	24
RR	6	4	12	10	17	19	19	24	20
SR	5	6	14	9	12	13	15	10	14
SC	12	14	17	21	19	20	25	17	28
SI	10	12	7	8	4	4	4	3	3

Table 5. The gain in frequency from using cardinal information

	3/3	3/15	6/6	6/12	9/9	12/6	12/12	15/3	15/15
ROC	3	−11	−10	−13	6	−30	−33	−31	−34
RS	36	51	70	80	78	72	85	49	88
RR	7	8	−49	−58	−80	−86	−103	−65	−107
SR	23	36	30	30	24	17	23	4	24
SC	27	38	63	76	75	66	78	48	86
SI	33	49	59	66	60	54	63	34	60

this spread for the three information situations ordinal, semi-cardinal, and cardinal as above. For completeness, the spread for the situation with one blank position is shown in Appendix A.

Table 6. The spread in frequency when using only ordinal information

	3/3	3/15	6/6	6/12	9/9	12/6	12/12	15/3	15/15
ROC	31	13	49	51	80	88	100	74	122
RS	10	51	76	83	108	127	144	107	168
RR	13	31	50	44	104	132	142	118	170
SR	4	41	12	26	13	7	13	0	11
SC	6	27	47	59	77	93	114	89	141
SI	10	51	76	83	108	127	144	107	168

Table 7. The spread in frequency when using semi-cardinal information

	3/3	3/15	6/6	6/12	9/9	12/6	12/12	15/3	15/15
ROC	23	5	40	42	69	80	94	66	117
RS	10	52	71	77	101	109	130	93	142
RR	0	40	16	7	62	95	104	90	133
SR	5	47	30	40	28	14	22	6	9
SC	2	32	43	55	66	77	91	71	110
SI	9	54	79	85	113	128	145	108	170

Table 8. The spread in frequency when using cardinal information

	3/3	3/15	6/6	6/12	9/9	12/6	12/12	15/3	15/15
ROC	39	60	64	72	133	80	87	54	110
RS	20	8	23	25	54	64	72	59	92
RR	27	18	54	54	75	92	88	83	115
SR	24	11	28	27	37	46	49	37	59
SC	26	22	4	0	18	33	39	37	54
SI	15	2	38	41	80	92	108	82	129

4.3 Selection Guidelines

Studying the results in Tables 1, 2, 3, 4, 5, 6, 7 and 8, the following conclusions can be drawn (which also make up the guidelines for selecting an automatic weight-generating function):

- The gain from using semi-cardinal information (allowing '=') is around 1–2% in decision power (Table 4), almost regardless of algorithm category and size of problem (the exception being SI (Simos) that gains less). This points to the viability of always allowing '=' ($>_0$ in our terminology) as a mode of expression, even in settings where there is usually strictly ordinal information. There is no compelling reason to use ordinal-only modes of expression.
- Up to 5–8% can be gained in decision power from allowing up to two blank positions in the rankings for some methods, but the choice of weight generation method is critical in this scenario. Since the generating functions are already at around 80% in their ordinal form, this is a very substantial gain overall.
- Algorithms from the category **N1** (RR and ROC) are, unsurprisingly, not able to handle cardinal information well. There is even a substantial *loss* in decision power when more information is used as opposed to the expected gain.

- The algorithm RR is not realistically usable for any purpose, be it ordinal or cardinal. Its performance is dominated by several others in every respect, and from the viewpoint of guidelines, it should be avoided.
- For really small problems, with only a few criteria and alternatives, any established algorithm will do (with the exception of equal weights, which is not covered in this article).
- The algorithm category **M** always outperforms the other categories. Next comes category **N** with some merit, while category **N1** is unsuitable for most purposes.
- Beyond two blank positions, there are diminishing returns. While not shown explicitly in the tables in this article, there cannot be much improvement per new blank position since there is simply no room for it (trying to approach an unrealistic 100% upper limit). Further, it is rather inconceivable that a decision-maker would be able to regularly discriminate between, say, five and six blank positions in a meaningful way.
- Within category **M**, in general the algorithm SR performs the best for ordinal information and SC (SUMROC) for cardinal information. But the picture is not as clear-cut. Considering the desired property of robustness as well (which SR was specifically designed for), the suggestion for a selection guideline would be to use SR if there is ordinal or semi-cardinal information, and SC if there is truly cardinal information (blanks present in the data). This way, the best performance of both weight-generating algorithms of this category can be obtained.
- Within category **N**, the algorithm RS performs better than SI (Simos) for cardinal and semi-cardinal information (for ordinal information, they are the same identical algorithm). This is somewhat surprising since SI – unlike RS – was designed for cardinal information, not the other way around. Nevertheless, it is suggested to look for an algorithm belonging to category **M** in the first place.

This set of observations and conclusions constitute the verified guidelines for selecting an automatic weight-generating function for MCDM applications, in ordinal as well as cardinal information settings. To sum up, if there is a considerable amount of cardinal information in the dataset considered, SUMROC is the best choice for generating weights.

5 Concluding Remarks

In this article, we have reviewed and measured a set of leading well-known automatic weight-generating functions that do not contain external parameters, i.e. they do not require any other information from a decision-maker than the ranking. We have also provided a set of guidelines for selecting automatic weight generation functions. One general observation is that there is a significant gain in decision power from allowing up to two strength positions (blanks) in the rankings, while the use of more positions only marginally improves the results. We have also seen that Rank Reciprocal and ROC handle cardinal information quite badly in their original format and that the performance of Rank Reversal is dominated by several others in virtually every respect, except in small problems where basically any algorithm will do. Furthermore, methods that are designed to work in scenarios with both N and N–1 degrees of decision-maker weight-selecting freedom always outperform methods from the other categories. We have moreover analysed a combined ranking method, SUMROC, for MCDM problems with respect to its performance compared to others and shown that if there is any truly cardinal information in the dataset considered, SUMROC is the best choice for generating surrogate weights.

Acknowledgements. This research was funded by the European Union's Horizon 2020 Programme call H2020-INFRAEOSC-05–2018-2019, Grant agreement number 831644, via the EOSCsecretariat.eu, and the EU project Co-Inform (Co-Creating Misinformation-Resilient Societies H2020-SC6-CO-CREATION-2017).

Appendix A

The Tables A1, A2 and A3 display the results for cardinal information with at most one blank position (symbols $>_0$, $>_1$, and $>_2$). As expected, the results fall between the semi-cardinal situation and the situation with at most two blank positions. All numbers in ‰.

Table A1. The cardinal frequency for the methods using a combination of DoFs

	3/3	3/15	6/6	6/12	9/9	12/6	12/12	15/3	15/15
ROC	899	793	828	794	797	808	773	852	756
RS	921	832	884	857	871	888	862	919	856
RR	899	789	779	735	702	693	641	736	594
SR	914	823	864	837	845	860	831	897	824
SC	920	830	888	863	877	899	873	925	867
SI	916	830	872	845	854	870	842	905	831

Table A2. The gain in frequency from using cardinal information

	3/3	3/15	6/6	6/12	9/9	12/6	12/12	15/3	15/15
ROC	11	10	5	5	−1	−10	−12	−16	−16
RS	33	45	49	55	54	52	58	35	61
RR	10	12	−25	−30	−44	−49	−61	−42	−65
SR	21	36	27	30	24	19	22	8	24
SC	27	37	42	54	47	51	57	34	62
SI	28	43	37	43	37	34	38	21	36

Table A3. The spread in frequency when using cardinal information

	3/3	3/15	6/6	6/12	9/9	12/6	12/12	15/3	15/15
ROC	35	48	65	71	80	91	112	71	126
RS	11	14	40	44	73	86	95	73	112
RR	21	1	49	47	77	99	113	90	132
SR	15	8	14	8	19	31	34	34	46
SC	22	1	10	18	29	50	58	52	77
SI	14	18	52	59	95	110	124	93	148

Appendix B

The Tables B1, B2, B3, B4, B5, B6, B7 and B8 display the results in % for the separate situations of *N*–1 and *N* DoF.

Table B1. The ordinal frequency for the methods using *N*–1 DoF

	3/3	3/15	6/6	6/12	9/9	12/6	12/12	15/3	15/15
ROC	904	790	848	815	838	862	835	905	833
RS	883	762	797	761	763	773	732	831	711
RR	896	762	829	787	798	808	773	837	744
SR	895	767	831	794	815	838	803	889	795
SC	896	780	823	780	792	802	759	847	735
SI	883	762	797	761	763	773	732	831	711

Table B2. The ordinal frequency for the methods using *N* DoF

	3/3	3/15	6/6	6/12	9/9	12/6	12/12	15/3	15/15
ROC	873	777	799	764	758	774	735	831	711
RS	893	813	873	844	871	900	876	938	879
RR	883	793	779	743	694	676	631	719	574
SR	891	808	843	820	828	845	816	889	806
SC	890	807	870	839	869	895	873	936	876
SI	893	813	873	844	871	900	876	938	879

Table B3. The semi-cardinal frequency for the methods using *N*–1 DoF

	3/3	3/15	6/6	6/12	9/9	12/6	12/12	15/3	15/15
ROC	915	808	864	834	852	872	849	911	846
RS	893	770	814	778	784	800	760	852	748
RR	895	761	824	779	794	809	773	847	746
SR	896	770	836	796	819	847	813	896	810
SC	906	791	842	803	816	830	796	873	778
SI	894	772	803	768	765	776	736	833	713

Table B4. The semi-cardinal frequency for the methods using *N* DoF

	3/3	3/15	6/6	6/12	9/9	12/6	12/12	15/3	15/15
ROC	892	803	824	792	783	792	755	845	729
RS	903	822	885	855	885	909	890	945	890
RR	895	801	808	772	732	714	669	757	613
SR	901	817	866	836	847	861	835	902	819
SC	904	823	885	858	882	907	887	944	888
SI	903	826	882	853	878	904	881	941	883

Table B5. The cardinal frequency (at most one blank) using *N*–1 DoF

	3/3	3/15	6/6	6/12	9/9	12/6	12/12	15/3	15/15
ROC	917	817	861	830	837	854	829	888	819
RS	927	825	864	835	835	845	815	883	800
RR	910	789	804	759	741	743	698	781	660
SR	922	819	871	841	855	876	848	914	847
SC	931	831	883	854	863	874	844	899	829
SI	923	821	846	816	807	815	780	859	757

Table B6. The cardinal frequency (at most one blank) using *N* DoF

	3/3	3/15	6/6	6/12	9/9	12/6	12/12	15/3	15/15
ROC	882	769	796	759	757	763	717	817	693
RS	916	839	904	879	908	931	910	956	912
RR	889	790	755	712	664	644	585	691	528
SR	907	827	857	833	836	845	814	880	801
SC	909	830	893	872	892	924	902	951	906
SI	909	839	898	875	902	925	904	952	905

Table B7. The cardinal frequency (at most two blanks) using *N*–1 DoF

	3/3	3/15	6/6	6/12	9/9	12/6	12/12	15/3	15/15
ROC	911	802	845	812	871	828	796	864	793
RS	934	842	894	870	868	876	853	904	837
RR	910	794	782	734	704	702	643	755	610
SR	928	829	881	851	864	881	857	912	854
SC	933	842	911	885	896	898	875	921	864
SI	929	837	875	848	837	844	813	877	791

Table B8. The cardinal frequency (at most two blanks) using *N* DoF

	3/3	3/15	6/6	6/12	9/9	12/6	12/12	15/3	15/15
ROC	872	742	781	740	738	748	709	810	683
RS	914	834	917	895	922	940	925	963	929
RR	883	776	728	680	629	610	555	672	495
SR	904	818	853	824	827	835	808	875	795
SC	907	820	907	885	914	931	914	958	918
SI	914	835	913	889	917	936	921	959	920

References

1. von Winterfeldt, D., Edwards, W.: Decision Analysis and Behavioural Research. Cambridge University Press (1986)
2. Barron, F., Barrett, B.: Decision quality using ranked attribute weights. Manag. Sci. **42**(11), 1515–1523 (1996)
3. Jia, J., Fischer, G.W., Dyer, J.: Attribute weighting methods and decision quality in the presence of response error: a simulation study. J. Behav. Decis. Making **11**(2), 85–105 (1998)
4. Barron, F., Barrett, B.: The efficacy of SMARTER: simple multi-attribute rating technique extended to ranking. Acta Psych. **93**(1–3), 23–36 (1996)
5. Katsikopoulos, K., Fasolo, B.: New tools for decision analysis. IEEE Trans. Syst. Man, Cyber. Part A: Syst. Hum. **36**(5), 960–967 (2006)
6. Stillwell, W., Seaver, D., Edwards, W.: A comparison of weight approximation techniques in multiattribute utility decision making. Org. Behavior Hum. Perform. **28**(1), 62–77 (1981)
7. Barron, F.H.: Selecting a best multiattribute alternative with partial information about attribute weights. Acta Psych. **80**(1–3), 91–103 (1992)
8. Danielson, M., Ekenberg, L.: Rank ordering methods for multi-criteria decisions. In: Zaraté, P., Kersten, G.E., Hernández, J.E. (eds.) Group Decision and Negotiation. A Process-Oriented View. GDN 2014. Lecture Notes in Business Information Processing, vol. 180. Springer, Cham (2014)

9. Danielson, M., Ekenberg, L.: Trade-offs for ordinal ranking methods in multi-criteria decisions. In: Bajwa, D., Koeszegi, S., Vetschera, R. (eds.) Group Decision and Negotiation. Theory, Empirical Evidence, and Application. GDN 2016. Lecture Notes in Business Information Processing, vol. 274. Springer, Cham (2017)
10. Simos, J.: L'evaluation environnementale: Un processus cognitif neegociee. Theese de doctorat, DGF-EPFL, Lausanne (1990)
11. Simos, J.: Evaluer l'impact sur l'environnement: Une approche originale par l'analyse multicriteere et la negociation. Presses Polytechniques et Universitaires Romandes, Lausanne (1990)
12. Danielson, M., Ekenberg, L.: A robustness study of state-of-the-art surrogate weights for MCDM. Group Decis. Negot. **26**(4), 677–691 (2016). https://doi.org/10.1007/s10726-016-9494-6
13. Danielson, M., Ekenberg, L.: An improvement to swing techniques for elicitation in MCDM methods. Knowl.-Based Syst. (2019). https://doi.org/10.1016/j.knosys.2019.01.001
14. Arbel, A., Vargas, L.G.: Preference simulation and preference programming: robustness issues in priority derivation. Eur. J. Oper. Res. **69**, 200–209 (1993)
15. Stewart, T.J.: Use of piecewise linear value functions in interactive multicriteria decision support: a monte carlo study. Manag. Sci. **39**(11), 1369–1381 (1993)
16. Ahn, B.S., Park, K.S.: Comparing methods for multiattribute decision making with ordinal weights. Comput. Oper. Res. **35**(5), 1660–1670 (2008)
17. Danielson, M., Ekenberg, L.: The CAR method for using preference strength in multi-criteria decision making. Group Decis. Negot. **25**(4), 775–797 (2015). https://doi.org/10.1007/s10726-015-9460-8
18. Rao, J.S., Sobel, M.: Incomplete Dirichlet integrals with applications to ordered uniform spacing. J. Multivar. Anal. **10**, 603–610 (1980)
19. Roberts, R., Goodwin, P.: Weight approximations in multi-attribute decision models. J. Multi-Criteria Decis. Anal. **11**, 291–303 (2002)

Efficiency of Competitiveness Evaluation of Medium-Lift Launch Vehicle (MLV) Using Integrated DEA-TOPSIS Model

Zhen Wan, Rustam Ismatov, and Haiyan Xu(✉)

College of Economics and Management, Nanjing University of Aeronautics and Astronautics, Nanjing 211106, Jiangsu, China
{wanzhen,roman,xuhaiyan}@nuaa.edu.cn

Abstract. Owing to increasing launch demand and lack of launch vehicle to space for commercial satellite customers, the commercial launch market for transporting satellites into orbit by medium-lift launch vehicle (MLV) is extremely popular and competitive. Understanding the efficiency and competitiveness of vehicles is particularly significant for both MLV providers and parties that are willing to utilize launch services. This paper presents a hybrid model that combines the data envelopment analysis (DEA) model, and technique for order performance by similarity to ideal solution (TOPSIS) to evaluate and then rank the efficiency and competitiveness of 15 currently operational MLVs. DEA analysis shows that the comprehensive efficiency, pure technical efficiency and scale efficiency of 10 MLVs are all 1, which means relatively DEA effective. The remaining 5 MLVs are relatively DEA ineffective, and the vehicle named Zenit 3 has the lowest relative efficiency. Through DEA, various criteria of ineffective MLVs are revised to ideal values, which achieves effectiveness and improve competitiveness. The following TOPSIS analysis shows that Falcon 9 Block 5 has the highest efficiency and competitiveness. This paper provides benchmark MLV with the highest efficiency, which can be studied extensively by the customers of commercial launch vehicles and peer MLV operators to compare different criteria affecting their efficiency and competitiveness.

Keywords: Medium-lift launch vehicle · Efficiency · Data envelopment analysis · Analytic hierarchy process · Technique for order performance by similarity to ideal solution

1 Introduction

1.1 Background

The era of space exploration and creation of various launch vehicles commenced with the first artificial satellite launch by Sputnik (SL-1) in 1957. Since then, increasingly more countries in the world have been trying to launch their own satellites into the orbit. Space technology and access to space have been elusive for developing countries [1].

A. T. de Almeida and D. C. Morais (Eds.): INSID 2021, LNBIP 435, pp. 36–52, 2021.
https://doi.org/10.1007/978-3-030-91768-5_3

Over the last half century, owing to the lack of awareness of policy/decision makers about the role of space technology in national development. Space technology was seen as very expensive and prestigious, meant only for the major industrialized countries, while the developing countries should focus on building their national economy and providing food, shelter and other social amenities for their ever-growing populations. In the last decade, the trend has changed, with many developing countries embracing space technology as one of the major ways of achieving sustainable development. This can be seen in countries like India, China, Morocco, South Africa, Nigeria, Egypt, and South Korea, among others [2].

The present trend towards small and medium sized satellites has also aided this transition because, apart from the smaller size, they tend to be cheaper to build and launch, with shorter development time, lower complexity, improved effectiveness and reduced operating costs [3]. This in turn has made them more affordable and has opened up now avenues for the acquisition of satellite technology.

However, a growing number of space players means an increasing demand for satellite launch vehicles that are able to carry payload to orbit.

1.2 Recent Development

Bangladesh has become the latest country to own a satellite. The Bangabandhu-1 satellite was launched on an improved version of SpaceX Falcon 9 rocket in 2018 [4]. Bangladesh's motivation for owning a satellite is strengthening the foundation for economic development, conforming to the growing trend of developing countries investing in their own satellites. However, the capacity of developing countries to generate enough revenue for recovering the satellite expenditure is questionable, especially if the local market is already crowded. The trend also comments on the validity of the United Nations and international space treaties stressing on sharing of space exploration benefits. The rising number of space actors also brings to forth the pressing need to ensure sustainability of outer space.

The World Bank has classified countries into low income, lower middle income, upper middle income and high-income sections. While none of the low-income countries own a satellite, a number of lower and upper middle-income economies have launched satellites. These satellites include communications, remote sensing and student satellites. The small satellites, based on the CubeSat standard, is allowing many countries to experiment with satellite building. These satellites are capable of carrying cameras and multispectral sensors for earth observation. India's PSLV has captured global attention for launching these satellites [5].

However, developing indigenous sophisticated remote sensing or communications satellites is beyond the scope of majority middle income countries and they are therefore dependent on established players. Venezuela, Pakistan, Nigeria, Cambodia, Laos, etc., have given contracts to China to launch their first communications satellites. Only a few such as India, China and Russia represent the middle-income group, with capacity to develop and launch satellites indigenously. India and China also offer cheaper launches compared to Western launchers [6].

Therefore, the aspiring countries are making efforts to indigenize satellite technology by partnering with established operators. For example, China is obliged to train the technicians from the ordering countries such as Venezuela, Bolivia, etc. [7]. However, a completely indigenous space program is not possible without mastering the launch technology. Given the associated military-strategic implications, the spacefaring countries will not be inclined to readily share or train on this technology. Moreover, a new launch capable country means increased competition and dwindling business opportunities to established operators.

Given the economic conditions, the middle-income countries also have to consider initial funding and operational expenses in addition to finding business opportunities. The contrasting stories of Nigeria and Belarus offer a good example in this scenario. Both launched their first communications satellites with China's loans and technical help. However, Nigeria had to compel its state entities to utilize the satellite before it becomes a liability. On the other hand, Belarus launched the satellite with business motive and has already leased half the capacity to users [8].

Bangladesh also leases some of the capacity to its neighbors. But it is already a crowded market. It is imperative to find business opportunities, given that satellites are costly and some of them are built on foreign loans. It is interesting that the middle-income countries have to obtain major loans for satellite services when the United Nations and the international space treaties emphasize helping the developing countries. It has been observed that a majority of Sustainable Development Goals can be achieved using satellites. The American GPS is free for use across the globe and also its Landsat earth observation data [9].

India has welcomed its South Asian neighbors to use its navigation satellite system free. However, it is hard to imagine communications satellites or earth observation images being offered free, across established players.

Still, it might be prudent to buy satellite data rather than launching one's own. It seems these countries have calculated that it is economical over the long term to use their own satellites rather than depend on external operators. The launching of own satellites also awards political mileage to incumbent leaders or help support a geopolitical narrative [10]. For example, Pakistan rejected the SAARC satellite and contracted China for its satellites. The African countries such as Ethiopia, Egypt, etc., can be seen using satellites as instruments for African leadership position.

While there could be more terrestrial issues behind the middle-income countries owning satellites, it should be noted that outer space is getting crowded. The competition for geostationary slots and frequency is well known. Tracking of satellites and mitigating threats from natural and man-made objects should be given priority during the planning stage. The increasing number of small satellite operators, concentrating on imagery and internet services, might offer a cheaper and secured option, given their ability to scale better than the countries in discussion.

It is welcoming to see that more middle-income countries are using satellite services in their economic development. A nuanced study for understanding the concurrent political, financial and space security implications, along with advances in business models across the globe, could be helpful in mitigating liabilities and fostering economic benefits.

1.3 A Medium-Lift Launch Vehicle (MLV)

A medium-lift launch vehicle (MLV) is the type of rocket system that can raise the satellite payload weighing between 2,000 and 20,000 kg to Low Earth Orbit (LEO), with altitude stretching from 200 km to 1600 km [5]. There are more than 100 MLVs in the world, yet majority are either already retired or under-development. At the time this research is carried out, the number of fully operational MLVs stated to be 15. For practicality purposes, the launch vehicles that are retired and/or under development were not analyzed during the study.

2 Literature Review

An overview of world space launches from 1957 through 1998, compared the launch system success rates among small, medium and large launch vehicles and space launch failures that occurred in all of the space-faring nations of the world. With particular attention paid to world space launch vehicle failures in 1983–1998, research found that former USSR, Europe, the US, Japan and China had between 6.2 and 17.2 failure rates, leading to the loss of launch vehicles and satellite payload estimated to be worth up to billions of US Dollars [1].

However, due to the high upfront costs of building a satellite and the potential long-term revenue stream such satellites can generate, satellite operators tend to emphasize launch reliability and schedule assurance over launch costs. Even though, as of 2010, the launch of a typical commercial GEO communications satellite can cost in excess of $100 million, it remains only a fraction of the overall cost of the system, taking into account the satellite manufacturing, insurance, and in many cases, ground equipment costs. While satellite operators are interested in lowering the cost of launch, that remains a lower priority than safely launching the satellite, and doing so on schedule, so that the satellite can generate revenue as soon as possible, and avoid other losses and liabilities that would be associated with the delayed entry into service of a new satellite [11].

Customers of the commercial launch vehicles also tend to prioritize other factors over launch cost, although they regard the launch cost to be very crucial aspect of their decision. Interestingly, A review of customer groups and their requirements for space launch offers that for most major customers, cost is not the most important, or even necessarily a key factor [12]. This flies in the face of conventional wisdom that suggests that lowering launch prices is paramount to national launch strategy. Most major government and established commercial customers prioritize other factors over cost. For exquisite national security space customers, launch reliability is critical because of the essential missions these spacecrafts perform, their high cost, and limited back-up options in the event of a failure. For civil human spaceflight, a related criterion, crew safety, is the key factor. Established commercial and civil robotic spacecraft customers do place some emphasis on cost, but place a higher priority on reliability and schedule assurance. A related issue that indicates that launch cost is not the key factor for most markets is the inelastic nature of most established government and commercial markets. Lowering launch prices does not necessarily stimulate additional demand for launches. This is in part because of launch costs are just one part of the overall cost of a system: for a commercial mission, the cost of launch may only be one-half to one-third or less of the total cost a new satellite system, given the large costs of satellite manufacturing, insurance, and other expenses. For government systems, launch costs may be even a smaller fraction for high-value national security or exploration missions that can run into the billions of dollars. Thus, even a significant drop in launch costs results in only a modest reduction in overall system costs. Based on the quarterly report of Federal Aviation Administration [13], the majority of the commercial satellite companies have used multiple types of launchers to deploy their satellites, indicating that launch procurers rarely confine themselves to a single launcher but prefer to diversify their choices. In doing so, a company makes decisions based on not one but many factors, evaluating them and making trade-offs to achieve an optimal combination of technical, programmatic, financial, and contractual factors.

A research revealed that two third of the SmallSats, the satellites with under 500 kg wet mass launched between 1995 and 2014 mainly used Medium-LVs, which offer reasonable launch prices, although may not offer the desired launch date and orbit. Besides taking medium size communication satellites to orbit, MLVs are also capable of carrying most SmallSats as secondary payload making them as optimal and universal choice for micro and mini-satellite developers [7]. Ability to carry small satellites is especially significant as small satellites are one of the most valuable tools to initiate and develop indigenous space capability. Apart from their low cost which makes them affordable by almost all countries, SmallSats offer the opportunity for developing countries to use the knowledge and skills of their citizens to develop their own satellites. These satellites offer low-cost data and accessibility via direct reception to low cost receiving stations, which can be operated by users [2].

3 Materials and Methods

3.1 Research Methodology

Data Envelopment Analysis (DEA)

Data Envelopment Analysis (DEA) is first proposed by operations researcher Cooper et al. It's a efficiency evaluation method based on relative efficiency. In this model, the evaluation object is regarded as a separate decision-making unit (DMU), the appropriate input and output indices are selected to construct a relatively efficient production frontier. According to the distance of each DMU and efficient production frontier, it can determine the relative effectiveness of each DMU [14].

The CCR model and the BCC model are two basic models that are frequently used. The CCR model assumes that the return to scale remains unchanged, and it measures comprehensive efficiency (CRSTE). The BCC model assumes that the return to scale is variable, and its measured efficiency includes pure technical efficiency (VRSTE) and scale efficiency (SCALE, SCALE = CRSTE/VRSTE) [15].

If the CRSTE value of a DMU is 1, the DMU is DEA effective. That means all inputs and outputs of the DMU is effective at the efficient production frontier. Or if it's less than 1, the DMU is not DEA effective. If the VRSTE value of a DMU is 1, the outputs of the DMU are the best at current situation. Or if it's less than 1, the optimal is not achieved and it needs to be improved. If the SCALE value of a DMU is 1, the DMU is in a state of constant return to scale. Or if it's less than 1, the scale of income is in ascending or descending state.

Analytic Hierarchy Process (AHP)
Analytic Hierarchy Process (AHP) which provides an effective method to deal with complex decision making, was first introduced by Thomas Saaty in 1970. It has been widely used in economic, social and management sciences [15].

The procedure of AHP is divided into five steps given below:

(1) Model the problem containing the decision goal, the alternatives for reaching it, and the criteria for evaluating the alternatives.
(2) Establish priorities among the elements of the hierarchy by making a series of judgments based on pairwise comparisons of the elements.
(3) Synthesize these judgments to yield a set of overall priorities for the hierarchy.
(4) Check the consistency of the judgments.
(5) Come to a final decision based on the results of this process.

Technique for Order Performance by Similarity to Ideal Solution (TOPSIS)
TOPSIS is a sorting method that is close to the ideal solution, which also known as the distance method of superior and inferior solutions, was first proposed by Hwang and Yoon in 1981 [16]. It is a comprehensive evaluation method, which often used in multi-objective decision analysis. The TOPSIS method ranks according to the closeness of a limited number of evaluation objects to the idealized object. It has no special requirements on sample size, only simple calculation and flexible application.

The basic principle is to sort by detecting the distance between the evaluation object and the optimal solution and the worst solution. If the evaluation object is closest to the optimal solution and farthest away from the worst solution, it is the best; otherwise, it is not optimal. Among them, each index value of the optimal solution reaches the optimal value of each evaluation index, each index value of the worst solution reaches the worst value of each evaluation index.

4 Results Analysis Based on the Integrated Model DEA-TOPSIS for MLVs

4.1 Results of DEA for MLVs

In this paper, 10 indices have been determined, including 2 output indices and 8 input indices. Selecting 15 vehicles as DMUs which come from Russia, India, China, America and so on, as shown in Table 1.

Table 1. DMU and index of DEA for MLVs.

DMU	Origin	Output		Input		
		Success rate	Pollution	Cost	Launches	Weight
Zenit-3	Ukraine	0.91	0.58	90.0	46	462200
Soyuz-2	Russia	0.95	0.89	48.5	112	312000
Soyuz-2-1v		0.83	0.88	38.5	6	158000
PSLV	India	0.94	0.69	23.0	53	281000
Long March 7	China	1.00	0.56	87.5	2	573000
Long March 3E		0.95	0.56	50.0	74	442385
Long March 3A		1.00	0.99	70.0	27	241000
Long March 2F		1.00	0.49	68.1	14	464000
Long March 2D		0.98	0.98	23.2	51	232250
H-IIA	Japan	0.98	1.00	90.0	43	365000
GSLV Mk. III	India	1.00	0.66	51.0	4	640000
GSLV Mk. II		0.86	0.77	47.0	7	414750
Falcon 9 Block 5	America	1.00	0.60	110.0	57	549054
Atlas V		0.99	0.45	110.0	85	590000
Antares 230		0.93	0.80	83.0	14	298000
DMU	Origin	Input				
		Height	Diameter	Max thrust	LEO	GTO
Zenit-3	Ukraine	0.91	0.58	90.0	46	462200

(continued)

Table 1. (*continued*)

DMU	Origin	Input				
		Height	Diameter	Max thrust	LEO	GTO
Soyuz-2	Russia	0.95	0.89	48.5	112	312000
Soyuz-2-1v		0.83	0.88	38.5	6	158000
PSLV	India	0.94	0.69	23.0	53	281000
1Long March 7	China	1.00	0.56	87.5	2	573000
Long March 3E		0.95	0.56	50.0	74	442385
Long March 3A		1.00	0.99	70.0	27	241000
Long March 2F		1.00	0.49	68.1	14	464000
Long March 2D		0.98	0.98	23.2	51	232250
H-IIA	Japan	0.98	1.00	90.0	43	365000
GSLV Mk. III	India	1.00	0.66	51.0	4	640000
GSLV Mk. II		0.86	0.77	47.0	7	414750
Falcon 9 Block 5	America	1.00	0.60	110.0	57	549054
Atlas V		0.99	0.45	110.0	85	590000
Antares 230		0.93	0.80	83.0	14	298000

DEAP2.1 software is used to analyze the efficiency of 15 MLVs. The calculation results shown on Table 2 and Fig. 1 highlight that the comprehensive efficiency of Soyuz-2, Soyuz-2-1v, PSLV, Long March 7, Long March 3A, Long March 2F, Long March 2D, GSLV Mk. III, GSLV Mk. II and Antares 230 reached an optimal level, where they are relatively effective. At the same time, the scale reward of these MLVs is unchanged. The comprehensive efficiency value of Zenit-3, Long March 3E, H-IIA, Falcon 9 Block 5 and Atlas V are less than 1, meaning that these vehicles are relatively ineffective and have diminishing returns for scale reward. Through DEA, each index of these DEA ineffective vehicles is analyzed and revised to the optimal values (Table 2), which can improve the comprehensive efficiency of these vehicles.

As shown in Table 3, in order to improve the comprehensive efficiency of Zenit-3, its pollution should be decreased by 0.353 tons with the condition that the input indices remain unchanged. When the output indices remain unchanged, the cost of per launch

Table 2. Efficiency of Each MLV.

DMU	CRSTE	VRSTE	SCALE	SCALE REWARD
Zenit-3	0.745	0.910	0.818	drs
Soyuz-2	1.000	1.000	1.000	–
Soyuz-2-1v	1.000	1.000	1.000	–
PSLV	1.000	1.000	1.000	–
Long March 7	1.000	1.000	1.000	–
Long March 3E	0.887	0.958	0.925	drs
Long March 3A	1.000	1.000	1.000	–
Long March 2F	1.000	1.000	1.000	–
Long March 2D	1.000	1.000	1.000	–
H-IIA	0.983	1.000	0.983	drs
GSLV Mk. III	1.000	1.000	1.000	–
GSLV Mk. II	1.000	1.000	1.000	–
Falcon 9 Block 5	0.828	1.000	0.828	drs
Atlas V	0.813	0.990	0.821	drs
Antares 230	1.000	1.000	1.000	–

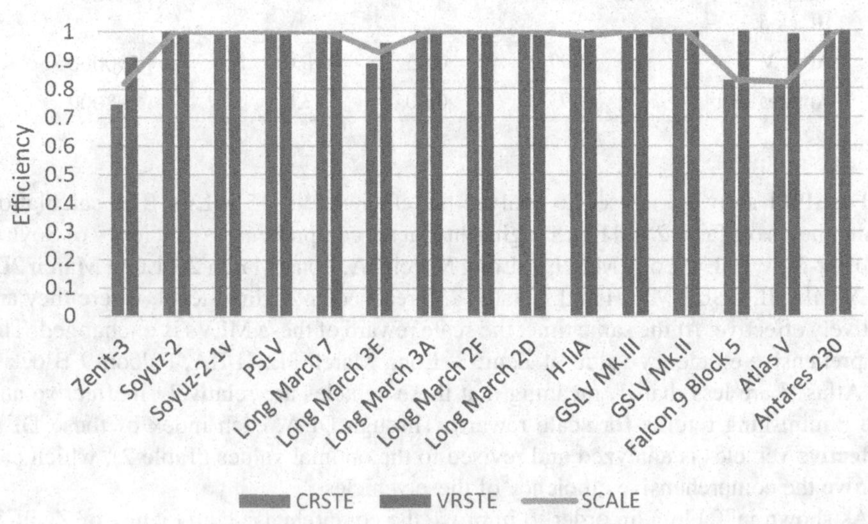

Fig. 1. Efficiency of MLVs.

Table 3. The projected value of DEA effective for MLVs.

DMU	Index		Original value	Radial movement	Slack movement	Projected value
Zenit-3	Output	Success rate	0.910	0.090	0.000	1.000
		Pollution	0.580	0.057	0.353	0.990
	Input	Cost	90.000	0.000	−20.000	70.000
		Launches	46.000	0.000	−19.000	27.000
		Weight	462200.000	0.000	−221200.000	241000.000
		Height	59.600	0.000	−7.080	52.520
		Diameter	3.900	0.000	−0.550	3.350
		Max thrust	8180.000	0.000	−5218.400	2961.600
		LEO	7000.000	0.000	−1000.000	6000.000
		GTO	6160.000	0.000	−3560.000	2600.000
Long March 3E	Output	Success rate	0.950	0.041	0.000	0.991
		Pollution	0.560	0.024	0.401	0.986
	Input	Cost	50.000	0.000	0.000	50.000
		Launches	74.000	0.000	−36.744	37.256
		Weight	442385.000	0.000	−205124.316	237260.684
		Height	55.500	0.000	−7.877	47.623
		Diameter	3.350	0.000	0.000	3.350
		Max thrust	2961.600	0.000	0.000	2961.600
		LEO	11500.000	0.000	−6568.376	4931.624
		GTO	5300.000	0.000	−3255.556	2044.444
Falcon9 Block 5	Output	Success rate	1.000	0.000	0.000	1.000
		Pollution	0.600	0.000	0.390	0.990
	Input	Cost	110.000	0.000	−40.000	70.000
		Launches	57.000	0.000	−30.000	27.000
		Weight	549054.000	0.000	−308054.000	241000.000
		Height	70.000	0.000	−17.480	52.520
		Diameter	3.660	0.000	−0.310	3.350
		Max thrust	7607.000	0.000	−4645.400	2961.600
		LEO	22800.000	0.000	−16800.000	6000.000

(continued)

Table 3. (*continued*)

DMU	Index		Original value	Radial movement	Slack movement	Projected value
		GTO	8300.000	0.000	−5700.000	2600.000
Atlas V	Output	Success rate	0.990	0.010	0.000	1.000
		Pollution	0.450	0.005	0.535	0.990
	Input	Cost	110.000	0.000	−40.000	70.000
		Launches	85.000	0.000	−58.000	27.000
		Weight	590000.000	0.000	−349000.000	241000.000
		Height	58.300	0.000	−5.780	52.520
		Diameter	3.810	0.000	−0.460	3.350
		Max thrust	3827.000	0.000	−865.400	2961.600
		LEO	18810.000	0.000	−12810.000	6000.000
		GTO	6800.000	0.000	−4200.000	2600.000

should be decreased by 20 million dollars, the launch times should be decreased by 19 times and so on.

4.2 Use AHP to Estimate the Weights of the Indices

The AHP method can estimate the weights of 10 indices which will be processed in TOPSIS. The structure of AHP includes 3 levels: objectives, criteria, and alternatives (Fig. 2). The objective in this paper is to evaluate the optimal vehicle, and the criteria are cost, launches, weight, height, diameter, max thrust, LEO, GEO, success rate and pollution. Alternatives are 15 vehicles.

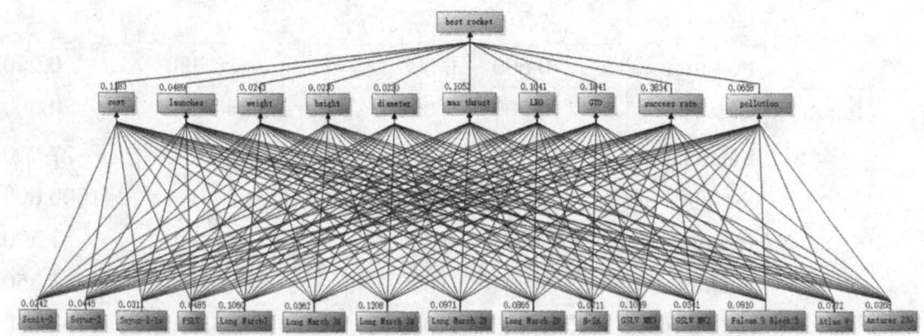

Fig. 2. Hierarchy structure for evaluating the criteria of MLVs.

Establishing priorities among the vehicles of the hierarchy is performed by making a series of judgments based on pairwise comparisons. To get the weight of each index, comparisons are paired using a scale of 1 to 9 and filled into a pair-wise comparison matrix, where each vehicle is compared with the other 14 vehicles. For example, according to 15 vehicles' success rate given in Table 1, the success rate of Soyuz-1 and Long March 3E are both 0.95. So the scale of pair-wise comparison of this two vehicles is 1, which represents Soyuz-1 and Long March 3E are equally important for success rate. Similarly, the success rate of Long March 7 is the highest while the success rate of GSLV MK. II is the lowest. Therefore, 9 represents Long March 7 is extremely important than GSLV MK. II for success rate. The implication of 1–9 scale is shown in Table 4. The weighted normalization matrix is shown in Table 5.

Table 4. The implication of 1–9 scale.

Scale	Implication
1	Two vehicles are equally important
3	The first vehicle is moderately important than the second vehicle
5	The first vehicle is quite important than the second vehicle
7	The first vehicle is demonstrably important than the second vehicle
9	The first vehicle is extremely important than the second vehicle

Table 5. Weighted normalization matrix for MLVs.

	Launches	Weight	Height	Diameter	Cost	Maxthrust	LEO	GTO	Success rate	Pollution
Launches	1.00	4.00	3.00	3.00	0.50	0.50	0.25	0.25	0.11	0.25
Weight	0.25	1.00	1.00	1.00	0.50	0.25	0.17	0.17	0.11	0.25
Height	0.33	1.00	1.00	1.00	0.33	0.25	0.17	0.17	0.11	0.25
Diameter	0.33	1.00	1.00	1.00	0.33	0.25	0.17	0.17	0.11	0.25
Cost	2.00	2.00	3.00	3.00	1.00	2.00	3.00	3.00	0.17	3.00
Maxthrust	2.00	4.00	4.00	4.00	0.50	1.00	2.00	2.00	0.17	4.00
LEO	4.00	6.00	6.00	6.00	0.33	0.50	1.00	1.00	0.17	3.00
GTO	4.00	6.00	6.00	6.00	0.33	0.50	1.00	1.00	0.17	3.00
Success rate	9.00	9.00	9.00	9.00	6.00	6.00	6.00	6.00	1.00	9.00
Pollution	4.00	4.00	4.00	4.00	0.33	0.25	0.33	0.33	0.11	1.00

C_i: 0.0980; weight to "optimal vehicle": 1.0000; λmax: 11.3136

The weight of these criteria is calculated in the AHP process, shown in Table 6.

Table 6. Weight of criteria for MLVs.

Index	Weight	Index	Weight
Cost	0.1183	Max thrust	0.1052
Launches	0.0489	LEO	0.1041
Weight	0.0243	GTO	0.1041
Height	0.0230	Success rate	0.3834
Diameter	0.0230	Pollution	0.0658

4.3 Results of TOPSIS

Let x_{ij} be the inputs for matrix of priorities where there are $i = 1, 2, \cdots, m$ alternatives and $j = 1, 2, \cdots, n$ criteria. There are five steps to implement TOPSIS method as follows:

$$X = \left(x_{ij}\right)_{10\times 15} = \begin{pmatrix} 90.0 & 46 & 462200 & 59.60 & 3.90 & 8180.0 & 7000 & 6160 & 0.91 & 0.58 \\ 48.5 & 112 & 312000 & 46.30 & 2.95 & 792.4 & 8200 & 3250 & 0.95 & 0.89 \\ 38.5 & 6 & 158000 & 44.00 & 3.00 & 1019.9 & 2800 & 1400 & 0.83 & 0.88 \\ 23.0 & 53 & 281000 & 44.00 & 2.80 & 4846.9 & 3800 & 1200 & 0.94 & 0.69 \\ 87.5 & 2 & 573000 & 60.13 & 3.35 & 2400.0 & 13500 & 7000 & 1.00 & 0.56 \\ 50.0 & 74 & 442385 & 55.50 & 3.35 & 2961.6 & 11500 & 5300 & 0.95 & 0.56 \\ 70.0 & 27 & 241000 & 52.52 & 3.35 & 2961.6 & 6000 & 2600 & 1.00 & 0.99 \\ 68.1 & 14 & 464000 & 62.00 & 3.35 & 3256.0 & 8400 & 3500 & 1.00 & 0.49 \\ 23.2 & 51 & 232250 & 41.06 & 3.35 & 2961.6 & 3500 & 1300 & 0.98 & 0.98 \\ 90.0 & 43 & 365000 & 53.00 & 4.00 & 1098.0 & 12500 & 5100 & 0.98 & 1.00 \\ 51.0 & 4 & 640000 & 43.43 & 4.00 & 5150.0 & 10000 & 4000 & 1.00 & 0.66 \\ 47.0 & 7 & 414750 & 49.13 & 2.80 & 4846.9 & 5000 & 2700 & 0.86 & 0.77 \\ 110.0 & 57 & 549054 & 70.00 & 3.66 & 7607.0 & 22800 & 8300 & 1.00 & 0.60 \\ 110.0 & 85 & 590000 & 58.30 & 3.81 & 3827.0 & 18810 & 6800 & 0.99 & 0.45 \\ 83.0 & 14 & 298000 & 41.90 & 3.90 & 3265.0 & 8000 & 3000 & 0.93 & 0.80 \end{pmatrix}$$

Step 1-Normalizing the original data

$$y_{ij} = \frac{x_{ij}}{\sqrt{\sum_{i=1}^{m} \sum_{j=1}^{n} x_{ij}^2}} \tag{1}$$

$$Y = (y_{ij})_{10\times 15} = \begin{pmatrix} 0.325 & 0.233 & 0.280 & 0.292 & 0.291 & 0.499 & 0.165 & 0.343 & 0.247 & 0.200 \\ 0.175 & 0.566 & 0.189 & 0.227 & 0.220 & 0.048 & 0.194 & 0.181 & 0.256 & 0.306 \\ 0.139 & 0.030 & 0.096 & 0.215 & 0.224 & 0.062 & 0.066 & 0.078 & 0.225 & 0.304 \\ 0.083 & 0.268 & 0.170 & 0.215 & 0.209 & 0.295 & 0.090 & 0.067 & 0.255 & 0.238 \\ 0.316 & 0.010 & 0.347 & 0.294 & 0.250 & 0.146 & 0.319 & 0.389 & 0.270 & 0.192 \\ 0.181 & 0.374 & 0.268 & 0.272 & 0.250 & 0.181 & 0.272 & 0.295 & 0.256 & 0.192 \\ 0.253 & 0.137 & 0.146 & 0.257 & 0.250 & 0.181 & 0.142 & 0.145 & 0.270 & 0.340 \\ 0.246 & 0.071 & 0.281 & 0.304 & 0.250 & 0.198 & 0.199 & 0.195 & 0.270 & 0.167 \\ 0.084 & 0.258 & 0.141 & 0.201 & 0.250 & 0.181 & 0.083 & 0.072 & 0.265 & 0.339 \\ 0.325 & 0.217 & 0.221 & 0.259 & 0.298 & 0.067 & 0.295 & 0.284 & 0.264 & 0.344 \\ 0.184 & 0.020 & 0.388 & 0.213 & 0.298 & 0.314 & 0.236 & 0.223 & 0.270 & 0.228 \\ 0.170 & 0.035 & 0.252 & 0.241 & 0.209 & 0.295 & 0.118 & 0.150 & 0.232 & 0.266 \\ 0.397 & 0.288 & 0.333 & 0.343 & 0.273 & 0.464 & 0.539 & 0.462 & 0.270 & 0.207 \\ 0.397 & 0.430 & 0.358 & 0.285 & 0.284 & 0.233 & 0.445 & 0.378 & 0.267 & 0.154 \\ 0.300 & 0.071 & 0.181 & 0.205 & 0.291 & 0.199 & 0.189 & 0.167 & 0.251 & 0.276 \end{pmatrix}$$

Step 2-Making the weight normalized matrix

$$v_{ij} = w_i y_{ij}, \quad i = 1, 2, \cdots, m; \; j = 1, 2, \cdots, n \tag{2}$$

$$w_i = (0.1183,\ 0.0489,\ 0.0243,\ 0.023,\ 0.023,\ 0.1052,\ 0.1041,\ 0.1041,\ 0.3834,\ 0.0658)$$

$$V = (v_{ij})_{10\times 15} = \begin{pmatrix} 0.038 & 0.011 & 0.007 & 0.007 & 0.007 & 0.052 & 0.017 & 0.036 & 0.095 & 0.013 \\ 0.021 & 0.028 & 0.005 & 0.005 & 0.005 & 0.005 & 0.020 & 0.019 & 0.098 & 0.020 \\ 0.016 & 0.001 & 0.002 & 0.005 & 0.005 & 0.007 & 0.007 & 0.008 & 0.086 & 0.020 \\ 0.010 & 0.013 & 0.004 & 0.005 & 0.005 & 0.031 & 0.009 & 0.007 & 0.098 & 0.016 \\ 0.037 & 0.000 & 0.008 & 0.007 & 0.006 & 0.015 & 0.033 & 0.041 & 0.104 & 0.013 \\ 0.021 & 0.018 & 0.007 & 0.006 & 0.006 & 0.019 & 0.028 & 0.031 & 0.098 & 0.013 \\ 0.030 & 0.007 & 0.004 & 0.006 & 0.006 & 0.019 & 0.015 & 0.015 & 0.104 & 0.022 \\ 0.029 & 0.003 & 0.007 & 0.007 & 0.006 & 0.021 & 0.021 & 0.020 & 0.104 & 0.011 \\ 0.010 & 0.013 & 0.003 & 0.005 & 0.006 & 0.019 & 0.009 & 0.008 & 0.102 & 0.022 \\ 0.038 & 0.011 & 0.005 & 0.006 & 0.007 & 0.007 & 0.031 & 0.030 & 0.101 & 0.023 \\ 0.022 & 0.001 & 0.009 & 0.005 & 0.007 & 0.033 & 0.025 & 0.023 & 0.104 & 0.015 \\ 0.020 & 0.002 & 0.006 & 0.006 & 0.005 & 0.031 & 0.012 & 0.016 & 0.089 & 0.017 \\ 0.047 & 0.014 & 0.008 & 0.008 & 0.006 & 0.049 & 0.056 & 0.048 & 0.104 & 0.014 \\ 0.047 & 0.021 & 0.009 & 0.007 & 0.007 & 0.025 & 0.046 & 0.039 & 0.102 & 0.010 \\ 0.035 & 0.003 & 0.004 & 0.005 & 0.007 & 0.021 & 0.020 & 0.017 & 0.096 & 0.018 \end{pmatrix}$$

Step 3-Determining the ideal solution and the negative-ideal solution

$$S^{+} = \{v_1^{+}, v_2^{+}, \cdots v_n^{+}\} = \{(\max_j v_{ij} | i \in I'), (\min_j \ v_{ij}) | i \in I''\} \quad (3)$$

$$S^{-} = \{v_1^{-}, v_2^{-}, \cdots v_n^{-}\} = \{(\min_j \ v_{ij} | i \in I'), (\max_j \ v_{ij}) | i \in I''\} \quad (4)$$

I' is associated with benefit criteria, and I'' is associated with cost criteria (Table 7).

Table 7. Determine the ideal solution and the negative-ideal solution for MLVs.

Criteria	S^{+}	S^{-}
Launches	0.010	0.047
Weight	0.028	0.000
Height	0.002	0.009
Diameter	0.008	0.005
Cost	0.005	0.007
Maxthrust	0.052	0.005
Leo	0.056	0.007
Gto	0.048	0.007
Success rate	0.104	0.086
Pollution	0.023	0.010

Step 4-Calculating the relative closeness to the ideal solution

$$D_i^{+} = \sqrt{\sum_{j=1}^{n} \left(v_j^{+} - v_{ij}\right)^2}, \quad i = 1, 2, \cdots, m \quad (5)$$

$$D_i^{-} = \sqrt{\sum_{j=1}^{n} \left(v_j^{-} - v_{ij}\right)^2}, \quad i = 1, 2, \cdots, m \quad (6)$$

$$C_i = \frac{D_i^{-}}{D_i^{+} + D_i^{-}}, \quad i = 1, 2, \cdots, m \quad (7)$$

Falcon 9 Block 5 with a value of 0.6623 is the optimal vehicle and Atlas V with a value of 0.5437 is the second optimal vehicle on the competitiveness evaluation. Antares 230 and Soyuz-2-1v are the worst vehicles based on the TOPSIS analysis (Table 8).

Table 8. Calculate the relative closeness for MLVs.

DMU	D_i^+	D_i^-	C_i	Rank
Zenit-3	0.0543	0.0588	0.5200	3
Soyuz-2	0.0675	0.0449	0.3992	8
Soyuz-2-1v	0.0847	0.0330	0.2804	15
PSLV	0.0681	0.0491	0.4191	7
Long March 7	0.0599	0.0483	0.4462	6
Long March 3E	0.0507	0.0484	0.4886	4
Long March 3A	0.0691	0.0337	0.3280	13
Long March 2F	0.0644	0.0355	0.3554	12
Long March 2D	0.0725	0.0462	0.3891	9
H-IIA	0.0647	0.0407	0.3862	10
GSLV Mk. III	0.0545	0.0481	0.4690	5
GSLV Mk. II	0.0669	0.0398	0.3730	11
Falcon 9 Block 5	0.0412	0.0808	0.6623	1
Atlas V	0.0508	0.0605	0.5437	2
Antares 230	0.0678	0.0292	0.3010	14

5 Conclusion

Based on the research outcome by extensive analysis of DEA and TOPSIS methodologies, it can be inferred that Falcon 9 is currently optimal launch vehicle by efficiency and competitiveness as it stands out to be closest to ideal solution with rounded coefficient of 0.66. Falcon 9 is a reusable, two-stage rocket designed and manufactured by SpaceX for the reliable and safe transport of people and payloads into Earth orbit and beyond. Falcon 9 is also the world's first orbital class reusable rocket. Reusability allows SpaceX, the manufacturer of Falcon family launch vehicles to re-fly the most expensive parts of the rocket, which in turn drives down the cost of space access. Falcon 9 is closely followed by Atlas V which is regarded to be the second-best alternative based on the research results, with the rounded value of 0.54 to conceptual optimal MLV. Atlas V operates with expendable launch system and it is the fifth major mid-size version in the Atlas rocket family. This MLV is currently operated by United Launch Alliance, a joint venture between Lockheed Martin and Boeing. Atlas V is also one of the major NASA launch vehicles and currently available for commercial use.

Acknowledgement. The fundings were provided by National Natural Science Foundation of China (Grant No. 71971115) and the Basic Scientific Research of Nanjing University of Aeronautics and Astronautics (NG2020004).

References

1. Chang, I.S.: Overview of world space launches. J. Propul. Power **16**(5), 853–866 (2000)
2. Chizea, F.D.: Small Satellites in Developing Countries—An Integral Part of National Development. Springer, Dordrecht (2002). https://doi.org/10.1007/978-94-017-3008-2_34
3. Argoun, M.B.: Recent design and utilization trends of small satellites in developing countries. Acta Astronaut. **71**(71), 119–128 (2012)
4. Federal Aviation Administration: FAA Aerospace Forecast: Fiscal Years 2019–2039 (2019). https://www.faa.gov/data_research/aviation/aerospace_forecasts/media/FY2019-39_FAA_Aerospace_Forecast.pdf
5. Crisp, N., Smith, K., Hollingsworth, P.: Small satellite launch to leo: a review of current and future launch systems. Trans. Jpn. Soc. Aeronaut. Space Sci. **12**(ists29), 1073–1079 (2014)
6. Haase, E.E.: Trends in the commercial launch services industry. In: AIP Conference Proceedings, vol. 552, pp. 638–643 (2001)
7. Federal Aviation Administration: Selecting a Launch Vehicle: What Factors Do Commercial Satellite Customers Consider? (2001). https://www.faa.gov/about/office_org/headquarters_offices/ast/media/q22001.pdf
8. Federal Aviation Administration: Commercial Space Transportation Forecasts (2015). https://brycetech.com/reports/report-documents/Commercial_Space_Transportation_Forecasts_2015.pdf
9. Andrews, J.: Spaceflight secondary payload system (SSPS) and SHERPA Tug-a new business model for secondary and hosted payloads. In: 26th Annual AIAA/USU Conference on Small Satellites, Logan, USA (2012)
10. Bojorquez, O.J., Jun, C.: Risk level analysis for Hazard area during commercial space launch. In: 2019 IEEE/AIAA 38th Digital Avionics Systems Conference, pp. 1–6 (2019)
11. Foust, J.: For Military Launch, Failure is Not an Option (2010). http://www.spacepolitics.com/2010/03/27/for-military-launch-failure-is-not-an-option. Accessed 27 Mar 2010
12. Foust, J.: Space launch capabilities and national strategy considerations. Astropolitics **8**(2–3), 175–193 (2010)
13. Wekerle, T., Filho, J.P., Costa, L., Trabasso, L.G.: Status and trends of smallsats and their launch vehicles—an up-to-date review. J. Aerosp. Technol. Manage. **9**(3), 269–286 (2017)
14. Man, D., Zhang, H.: The study of DEA Application in tourism city: a case for members of the world tourism city federation in China. In: Zhang, Z., Shen, Z.M., Zhang, J., Zhang, R. (eds.) LISS 2014, pp. 831–835. Springer, Heidelberg (2015). https://doi.org/10.1007/978-3-662-43871-8_119
15. Liu, X., Gong, D., Liu, S.: Application of DEA model with AHP restraint cone in technical evaluation for road network. In: Zhang, Z., Shen, Z.M., Zhang, J., Zhang, R. (eds.) LISS 2014, pp. 717–723. Springer, Heidelberg (2015). https://doi.org/10.1007/978-3-662-43871-8_103
16. Wang, Y., Kong, H., Liu, F., Qu, Y.: Application of TOPSIS method on evaluating campus emergency management capacity. In: Zhang, Z., Zhang, R., Zhang, J. (eds.) 2nd International Conference on Logistics, Informatics and Service Science, pp. 1445–1450. Springer, Beijing (2013). https://doi.org/10.1007/978-3-642-32054-5_204

Japanese Port Alliance: Cooperative Game Theory Analysis

Rintaro Rai(✉), Sinndy Dayana Rico Lugo(✉), Nariaki Nishino, and Tomoya Kawasaki

The University of Tokyo, Tokyo, Japan
{r.rai,sinndydayana.rico}@css.t.u-tokyo.ac.jp

Abstract. As the volume of containerized cargo handled around the world continues to in-crease, ports and harbors are playing an increasingly important role in both economic and logistical terms. Japan is no exception to this trend. Its ports of Tokyo, Yokohama, and Osaka serve in important roles in Japan's logistics and international trade. Nevertheless, Japanese ports are currently unable to keep up with global container demand growth efficiently. Therefore, as described herein, based on cooperative game theory, the current situation of Japanese ports is studied, especially in terms of the international container strategic port policy begun in 2010 and alliances among ports. A study case was developed based on direct interviews, which, complemented by reviewed literature, reveals several facts related to the possibility for alliances among Japanese ports. Cooperative game theory was applied to elucidate the cores for three players in Tokyo Bay: Port of Tokyo, Port of Yokohama, and Port of Kawasaki. Finally, the study case results and the formulation of analysis as a cooperative game are discussed.

Keywords: Cooperative game · Japanese port · Port alliance

1 Introduction

As the volume of containerized cargo handled around the world continues to increase, ports and harbors are playing increasingly important roles in both economic and logistical terms. Japan, no exception to this trend, has the ports of Tokyo, Yokohama, and Osaka all playing salient roles in Japan's logistics network.

As described herein, the current situation of Japanese ports is examined based on game theory, especially the international container strategic port policy begun in 2010 and alliances among ports. Inoue [2] reported for Japanese ports that Osaka Bay is not actually operating as an integrated system. The study objectives are to conduct interviews with port personnel, mainly at Tokyo Bay, and to summarize the responses to interviews and literature references to analyze the current situations of ports. Next, the paper presents an examination of why alliances are not taking place and how alliances can be achieved, especially from a cooperative game theory approach.

A. T. de Almeida and D. C. Morais (Eds.): INSID 2021, LNBIP 435, pp. 53–67, 2021.
https://doi.org/10.1007/978-3-030-91768-5_4

2 Context of Japanese Ports

2.1 Current Japanese Port Situation

Japanese ports play an important role in the international trade of a country. However, currently, Japanese ports are unable to keep pace with the global container demand growth. On the one hand, Japan has been losing hub port functions [3]. On the other hand, Japan has been unable to handle Panamax-standard container ships from the early stages, which has caused a decrease in its number of trunk routes. Another effect of this situation is that the number of containers has not increased to the same degree as the rest of the world [4].

Another issue is that foreign global terminal operators have not yet entered the market or are unable to do so. In the case of Japan, a license must be obtained for each port to become a terminal operator. When the Super Hub Port, which was a policy of Japanese Government, was established in 2001, one aim was the development of domestic terminal operators. However, as a result, not even one nationwide terminal operator was developed because the integrated terminal operator (called a Mega Terminal operator) is a form of equal joint investments by participating port operators, which made it difficult to conduct unified management [5]. All of these policies failed to produce adequate results. The central government eventually admitted that the policy had failed [3].

Additionally, port management by local government, which is neither state-run nor private, is also an issue [2]. Consequently, no progress has occurred in large scale development involving surrounding areas because the administrators of, for example, the ports and the administrators of the roads and airports in the surrounding areas differ. Ports can develop independently, but for other areas, they must consult with the respective administrators.

2.2 International Container Strategic Port Policy

Therefore, in 2011, the Ministry of Land, Infrastructure, Transport, and Tourism of Japan (MLIT) formulated the International Container Strategic Port Policy. Under Democratic Party of Japan administration, the policy of "selection and concentration" had been pushed forward, aimed at maintaining and regaining trunk routes of Japan. Particularly Busan, South Korea, has been recognized as the strongest competitor of Japanese ports [6]. Under this policy, Tokyo Bay and Osaka Bay were selected as Strategic International Ports. The privatization of ports was simultaneously promoted. In Tokyo Bay, the Yokohama–Kawasaki International Port Corporation (YKIP) has been established and in Osaka Bay, Kobe–Osaka International Port Corporation (KOIP) has been established. Additionally, Nagoya and Yokkaichi have been selected as Central International Ports, which are quasi-strategic ports. Also, Nagoya and Yokkaichi International Port Corporation (NYP) has been established.

Strategic International Ports are defined as the highest level of ports, receiving a subsidy of from the government for direct quay wall construction [7]. In addition, the construction area can be leased to the private sector. In this sense, this paper presents a discussion of port alliance characteristics.

Figure 1 presents geographic locations of the main Japanese ports. They are Kobe, Osaka, Yokkaichi, Nagoya, Yokohama, Kawasaki, and Tokyo. All have been selected as Strategic International Ports or Central International Ports.

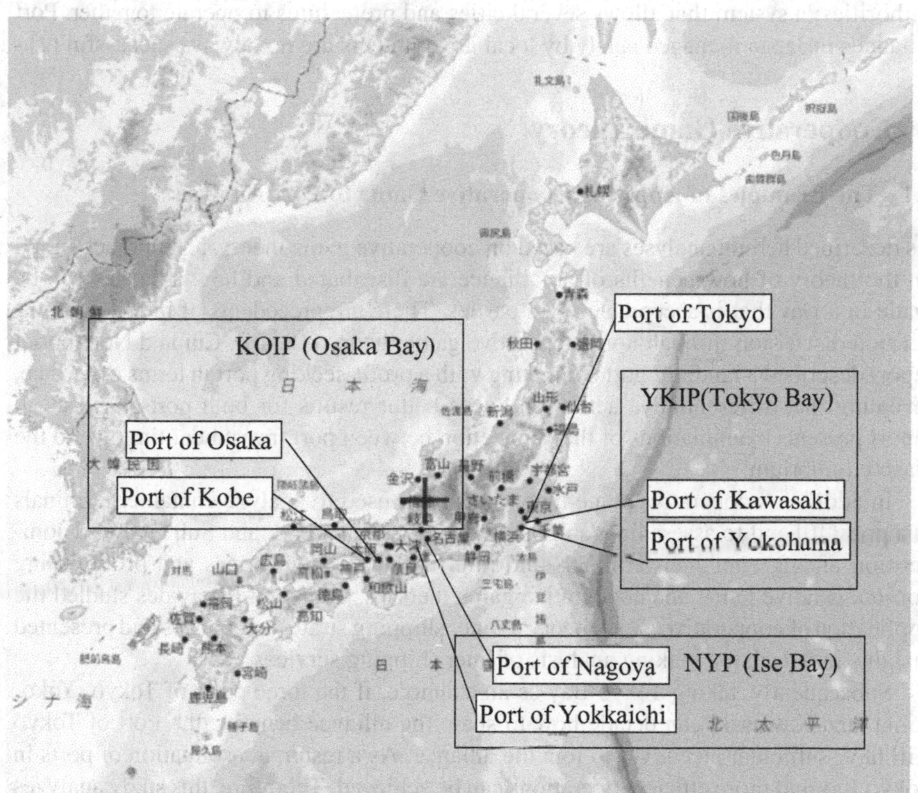

Fig. 1. Map of Japanese ports, with names of port based on the Geospatial Information Authority of Japan (Adapted by the authors from Geospatial Information Authority of Japan [8]).

Actually, KOIP has the largest share of government investment. Government is the largest shareholder, with 33.4%, corresponding to 500 million JPY (4.5 million USD) of capital: more than one-third of the total. The cities of Kobe and Osaka, which are the port administrators of the respective ports, also have stakes of nearly 1/3, equal to 450 million JPY (4.1 million USD). This capital distribution makes the decision-making process difficult and operations inefficient [9].

One Tokyo Bay joint-stock company is YKIP. The Tokyo Bay ports are designated by MLIT as the Keihin Port of Tokyo, the Keihin Port of Yokohama, and The Keihin Port of Kawasaki. A plan to merge the three ports, Tokyo, Yokohama, and Kawasaki existed at one time. However, immediately before that merger, the Port of Tokyo chose not to join the alliance, resulting in the formation of YKIP, an alliance between Yokohama and Kawasaki only. A main theme of the present study is therefore why the Port of Tokyo chose not to join the alliance. Those reasons are presented herein.

Although not selected as a Strategic International Port, the Port of Nagoya and the Port of Yokkaichi are Central International Ports; NYP, a joint venture between Nagoya and Yokkaichi, has been established. The Port of Nagoya and the Port of Yokkaichi differ from the ports of Tokyo Bay and Osaka Bay because they are operated by special district authorities: a system that allows several cities and prefectures to operate together. Port alliances in Japan managed solely by local governments are not always successful [2].

3 Cooperative Game Theory

3.1 The Examples of Applying Cooperative Game to Port Analysis

As described herein, analyses are based on cooperative game theory, an approach based on the theory of how benefits of an alliance are distributed and how alliances can be made in a way that is acceptable to all parties. There are precedents of inter-port competition discussion through non-cooperative game theory [18, 19]. Cui and Notteboom report described a landlord port competing with a profit-seeking port in terms of service, revealing that differentiative action produces better results for both ports. Ishii et al. report presents examinations of the connection between port investment timing and the Nash equilibrium.

In addition, cooperative game theory has been used to analyze container terminals at a port [10] and to assess liner shipping operations [11]. Park and Suh analyzed competition among container terminals and inferred equilibrium prices and profits using non-cooperative game and cooperative game theories. Song and Panayides studied the application of cooperative game theory to liner shipping strategic alliances and presented insights into decision making analysis of liner shipping services.

Subsequently, taking Tokyo Bay as an example, if the three ports of Tokyo, Yokohama, and Kawasaki can decide how to share the alliance benefits, the Port of Tokyo will have sufficient incentives to join the alliance. As a result, consolidation of ports in Tokyo Bay and more efficient operations can be achieved. Therefore, this study analyzes the three ports of Tokyo Bay as three players of a cooperative game. As the integration of Japanese ports progresses in the future, the research can be extended to diverse applications, including integrated operations of multiple ports.

3.2 Mathematical Formulation

Cooperative game theory is a different system of theory from commonly known non-cooperative game theory such as Nash equilibrium seeking. Non-cooperative games reveal what action is the best for a certain situation. In contrast, cooperative game theory was born from a book by von Neumann and Morgenstern [12], presenting an examination of what players can achieve through cooperation. Some applications of cooperative game theory to the field of logistics have already been described. For instance, Nagarajan et al. [13] reported that cooperative game theory gives a distribution of profit and sustainability for the supply chain. Cachon and Netessine [14] provided numerous related examples.

Subsequently, the basis of the cooperative game theory is presented above according to Nakayama et al. [15]. In cooperative game theory, a cooperative game includes a set of players, where k_i represents a player.

$$N = (k_1, k_2, \ldots k_n) \tag{1}$$

A set of characteristic functions $v(N, v)$ is used, where v is a characteristic function that yields a gain for every subset of players in $S(\subseteq N)$. Also, v is a function that returns 2^n possible values. For $\varnothing$ (empty set), it returns 0 ($v(\varnothing) = 0$) Here, in general, the characteristic function v is given super-additivity. That super-additivity means that, for any two sets S and T,

$$v(S) + v(T) \leq v(S \cup T) \tag{2}$$

Generally speaking, the objective of cooperative games is to ascertain how to distribute the benefits from cooperation. From the super-additive property, the overall gain will be higher if all members cooperate. Therefore, the emphasis is on how to formulate the imputation to each player when all players are in alliance. Here, the imputation vector is shown below.

$$\boldsymbol{x} = (x_1, x_2, \ldots, x_n) \tag{3}$$

The imputation x satisfies the following condition.

$$\sum_{i \in N} x_i = v(N) \tag{4}$$

$$x_i \geq v(\{k_i\}) \quad \forall k_i \in N \tag{5}$$

The first equation above implies total rationality. Therefore, the overall gain must be distributed without excess. The second equation represents individual rationality because there is no incentive to deviate from the alliance and form another alliance with the player alone. Next, the concept of the core is explained because it is important in cooperative game theory. The core is an imputation (i.e., a set of imputations) in a coalition game such that no imputation belonging to the set is dominated by some other imputation. Therefore, at its core, for all players, there is no other imputation which dominates the imputations in the core set. Defining the set of cores as x, we have the following.

$$\left\{ \boldsymbol{x} \middle| \sum\nolimits_{i \in S} x_i \geq v(S) \quad \forall S \subseteq N, S \neq N, \emptyset \right\} \tag{6}$$

4 Research Methodology

During the study, interviews were conducted with the companies involved. The study gathered perceptions and opinions of people in the field, which are not available in public documents. This feature made the research more meaningful and enhanced its contribution to the literature on this subject.

The information was collected by sending questions to each organization in advance via e-mail and by soliciting responses in an interview format. Consequently, interviews were conducted during April and May of 2021. During the period, the Tokyo Metropolitan Government and other prefectures of Japan had issued priority measures to prevent the spread of the COVID-19 disease and had declared a state of emergency.

This study is an explorative one. For this reason, interviewees, as a data collection instrument, were selected because of the convenience in obtaining information about the perspectives of the port staff about cooperativeness and also considering the extension of the COVID-19 restrictions. This time, the study specifically addressed the case of Tokyo Bay for convenience. Three major ports selected as Strategic International Ports are there. However, no opportunity existed to visit the locations directly because of public restrictions, therefore, interviews were conducted online.

Based on the interview data and literature references, the profits gained when ports mutually cooperate were calculated. Next, game formulation was conducted based on the profit.

Figure 2 presents the research process by showing the main steps.

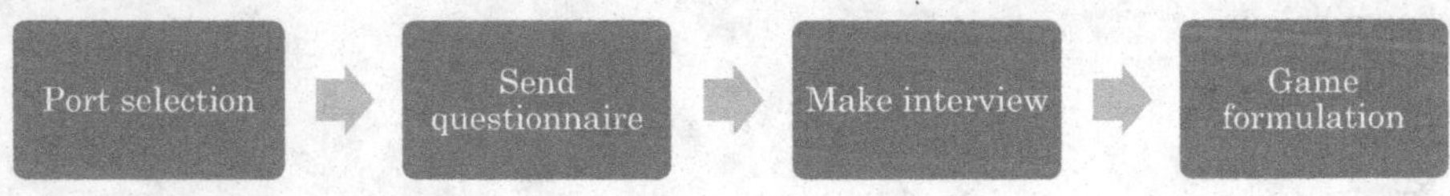

Fig. 2. Research flow.

Interviews proceeded according to the questionnaire sent in advance. The questions were shown below (Table 1).

Table 1. Question list

Questions
What are current status of the alliance and possible benefits?
What are obstacles to alliances of ports generally?
Why did not Port of Tokyo join the alliance? What are merits and demerits of it?
What do you think of Port of Busan? How important are trunk routes?
What is the situation of governances of joint port company based on investment ratio?

The selection of interviewees was based on the importance of the players and the convenience in contacting them. The first interviewee was the national governmental institution: the leader of port policy. The interviewee was asked about views on policy in general and perceptions of each port. Other interviewees were from major Japanese ports, although names are not disclosed here because of information management restrictions. Interviews of port governance professionals were also conducted.

5 The Result of Interview

In this section, the facts realized based on the conducted interviews and the related literature review are summarized.

5.1 Current Situation of Alliances of each Joint Venture

The first is the current situation of alliances created in the context of the International Container Strategic Port Policy. This section specifically examines the collaboration of the ports in proximity: the main topic of this study. Results show that no specific alliance exists among the joint venture for logistical optimization. In the first place, the main purpose of this policy is to accept calls of a large container vessel (i.e., trunk lines) at Japanese ports to compete with the growth of ports in Korea and China. Huge investments were made by the central government within this policy framework. To procure central government funds, the establishment of joint ventures in the form of alliances among ports was required by the central government. Some exchange of ideas occurred among the ports, but nothing beyond that, such as cargo exchange or terminal exchange.

5.2 Obstacles for the Alliance

The alliance has not worked efficiently. The first reason is the terminal operator licensing system. Currently, terminal operators are allowed to handle cargo only if they have a license that is specified for each port and terminal. This license makes it difficult for new operators (e.g., Global Terminal Operator) to enter the market. Also, moving terminals involves obtaining a new license. Under this situation, it would be difficult to move terminal operators among ports to specify which ports handle what type of cargo because it would require the reassembly of companies operating there.

The next issue is the Japanese Port Law. Currently, Japan's Port Law stipulates that the local government be the main port administrator. Therefore, other entities, such as private companies or other local governments, the central government, have difficulty being involved in operations. For a two-port partnership, it is unavoidable that both parties be involved in the operation of each port of the other. Therefore, the Port Act might represent an obstacle.

It turns out that an attempt to solve this difficulty was made before the International Container Strategic Port Policy. Today, Port of Nagoya and Port of Yokkaichi are each managed by the city and prefecture. In this sense, a special district authority is a form of administration in which the city and prefecture administrations are integrated. Applying this system, a plan existed to create a special district authority in Nagoya and Yokkaichi to operate the union through wide-area cooperation. It was not achieved. Even if attempted again, it would be difficult because of conflict between port transporters and cargo owners.

Another issue raised was that of ownership. For Tokyo Bay, ownership can be an obstacle to the current collaboration. In the case of Port of Tokyo, some facilities are owned by the Tokyo Metropolitan Government's Port and Harbor Bureau, but most facilities are owned by Tokyo Pier Corporation, a subsidiary of Tokyo Rinkai (HD). In the case of Port Yokohama, however, most of the property belongs to the central government or the city of Yokohama. In this situation, if a private joint venture is established, the question of how to handle the property ownership might become a relevant issue. Particularly on the Port of Tokyo side, it is thought that there will be considerable hurdles to the realization of the transfer of assets because of conflict with investors on the HD side.

5.3 Port of Tokyo

The fact that Port of Tokyo did not join the alliance in the earlier attempt is an additional inconvenience. The government of Japan regrets that Port of Tokyo, the largest port in Japan, did not join the alliance. It is still waiting for joining the alliance so that three Strategic International Ports in Tokyo Bay can conduct joint operations including port sales. The port operators belonging to YKIP also hope that the addition of Port of Tokyo will engender efficient sharing of port functions as a hub, for trans-shipment, and so forth. For instance, according to MLIT [16], Port of Tokyo will continue to serve as a hub for trunk routes and Asian routes, and as a gateway port of the hinterland including the Tokyo metropolitan area and northern Kanto area, which is a global consumer area, whereas Port of Yokohama will accept ultra-large container ships (e.g. 23,000 TEU[1]) and serve as a trans-shipment port by building berths with a depth greater than 18 m. Regarding the Port of Kawasaki, it is expected to share responsibility as a base for importing cargo from Asia using refrigerated and frozen warehouses in the area behind the port.

Port of Tokyo has not opposed to International Container Strategic Port Policy, to acquire trunk routes. Port of Tokyo did not join the alliance for several reasons. The first is the difference in port policies. Whereas Port of Yokohama and Port of Kawasaki are subsidized by the central government to invest in port infrastructure and collect both hinterland and trans-shipment cargo, the main objective of Port of Tokyo is to meet the demand generated and attracted in their hinterland as a gateway port. It is noteworthy that the hinterland cargo of Tokyo port is high and continuing to grow. In this sense, it does not find a need to join the alliance right now.

The second issue is central government involvement. Port of Tokyo itself seems to follow a general division of labor, as described in reference materials. Immediately after the International Container Strategic Port Policy was formulated, the three ports discussed the possibility of forming an alliance. However, around 2014, it emerged that the central government would be the main investor of a port management company, which raised concerns that alliance functions which the central government might propose would be disadvantageous to Port of Tokyo.

As a result, Port of Tokyo withdrew from the alliance. Port of Tokyo itself has not given up on the alliance, but its entry into the alliance in its current form is difficult.

[1] Twenty-foot Equivalent Unit.

For example, it is preferable for Port of Tokyo that the central government involves port management only by offering subsidies for each port infrastructure project, but not for the operation of the port management company. Under such circumstances, Port of Tokyo would consider joining the alliance.

5.4 Differences of the Attitude to Busan Port

Additionally, regarding Busan Port, the perspectives of the ports were obtained. The central government recognizes the ports of South Korea and China as competitors. For that reason, it is making a guided policy. In fact, the central government believes that each port shares this perception. Therefore, they are able to cooperate with each port. The same is true for Port of Yokohama, with the only difference that, as a practical matter, Busan Port is too large compared to the studied ports to regard it as a competitor.

In addition, the Port of Tokyo does not play a trans-shipment role as Busan Port does. In that sense, it is not regarded as a competitor. However, both ports share a sense of crisis: if things continue as they are, then their trunk routes will flow to Korean ports. Japan and its ports will lose their competitiveness.

5.5 Governance of the Alliance

The final insight obtained from the interviews and literature review is related to governance of the alliance. The national government is the largest shareholder of the joint ventures in Tokyo Bay and Osaka Bay, they were designated as Strategic International Ports by the central government. As described previously, the Port Law stipulates that local governments are mainly responsible for port management whereas the position of the central government is limited to providing necessary information, guidance, and advice, even after the International Container Strategic Port Policy.

In contrast, for YKIP, the main port management policy is set by the central government. Consequently, the role of the ports is to follow the guidelines of the central government and coordinate with the companies and businesses for the use of the ports. Particularly, Port of Yokohama has a dock, which was built recently with governmental investment. It is under the direct control of the central government. Regarding Ise Bay, the central government has not invested in the port. It has not been involved directly in port management. However, Ise Bay was also established as a Core International hub port. The merger took place because the Port of Nagoya needed government support. Because NYP adopts class shares, every share of stock of NYP is defined as Nagoya's right of management, all of which are possessed by the Port of Nagoya, Yokkaichi's right of management, almost all of which is possessed by the port of Yokkaichi and each general meeting of shareholders takes place differently. Therefore, the Nagoya and Yokkaichi ports operate almost separately. Such discussions are held mainly with the central government rather than between the ports.

5.6 The Importance of Current Users

Results showed that all ports consider shippers in their own hinterland as the most important factor. For example, in terms of management policy, emphasis is not on optimizing convenience of the port itself, but on whether or not it will benefit users. Regarding other benefits and shortcomings of cooperation, the first consideration is whether it will benefit current customers. This consideration is regarded as unavoidable at a certain level because each company is a customer. However, the stronger this stance is, the more the port operation itself is at risk of falling into localized optimization when viewed from a broader perspective.

6 Formulation as a Cooperative Game

In this chapter, the cooperative game theory is used to find the core for the case of three players in Tokyo Bay: Port of Tokyo, Port of Yokohama, and Port of Kawasaki.

Assuming that player names are k_1 for Port of Tokyo, k_2 for Port of Yokohama, and k_3 for Port of Kawasaki, the set of players is the following.

$$N = (k_1, k_2, k_3) \tag{7}$$

First, the gain that each player can achieve independently should be considered. This can be done by considering the values of 2015, which is the stage before cooperation. In this case, the number of containers will be examined specifically for several reasons. First, this paper is studying the alliance in the international "container" strategic port policy. Second, there is a positive correlation between container handling volume and port size [1]. Lastly, the economics of scale works very well in maritime logistics, and as the volume of containers, the marginal cost per container decreases. For these reasons, the number of container is best as indicators and profits of ports.

The unit is 10,000 TEU. The cargo volume handled in 2015 [20–22] before the establishment of YKIP is the gain achieved in a case of no cooperation.

$$v((k_1)) = 415 \tag{8}$$

$$v((k_2)) = 278 \tag{9}$$

$$v((k_3)) = 11 \tag{10}$$

Next, the gains earned if each port cooperates with the other will be examined. The profit of YKIP, which is the current alliance, is the total cargo volume of the ports of Yokohama and Kawasaki in 2019 because, at the time of research, only preliminary figures for cargo volumes in 2020 were available, and also because the situation in 2020 differed from a usual one because of the influences of the COVID19 epidemic, making comparison difficult.

Full-scale operations of the MC-4 terminal at Minami-Honmoku Pier, which can accommodate 400 m class container ships, will begin in 2021. This construction work is included among the benefits of cooperation because it is covered by subsidies. Nevertheless, the specific effects are not included because it is not in full operation.

Next, the ports of Tokyo and Yokohama, and the ports of Tokyo and Kawasaki are examined. Presumably, the Port of Tokyo will join the alliance if it receives support for the construction of the Y2 and Y3 terminals, which it considers as a gain from cooperation. According to data from Port of Tokyo, the volumes of cargo handled at the Y2 and Y3 terminals are estimated as 780,000 TEU [17]. Of this amount, 70%, which is the subsidized rate, is regarded as the alliance's gain; 550,000 TEU is added. The Y3 terminal has not been completed.

Then, each profit of the two player alliance is shown by combining cargo volumes of 2019 [20–22] and added 550,000 TEU only for Port of Tokyo.

$$v(k_1, k_2) = 822 \tag{11}$$

$$v(k_2, k_3) = 315 \tag{12}$$

$$v(k_3, k_1) = 573 \tag{13}$$

In Table 2, each amount of cargos is reported. The gains which are shown above are the combined figures in this table.

Table 2. Table of each container handling volume (10,000 TEU).

Ports\Year	2015	2019
Port of Tokyo	415	485 (Real), 540 (If Port of Tokyo joins YKIP)
Port of Yokohama	278	282
Port of Kawasaki	11	33

Finally, in the case of a joint operation of the three ports, the interviews revealed that the Port of Yokohama and the Port of Kawasaki are not currently operating efficiently or exchanging cargo. In the case of the Port of Tokyo, too, the gain is only attributable to subsidies. Even if the number of cooperating ports increases, the increase in gain will not change. Therefore, for joint operation of the three ports, the cargo volumes handled by each port can be simply added up. Therefore, the results are the following.

$$v(k_1, k_2, k_3) = 855 \tag{14}$$

Finding the core from the equation shown in Sect. 3, there is only one imputation that is the core.

$$x_{core} = (540, 282, 33) \tag{15}$$

The reason there is only one core imputation is that Port of Tokyo itself has calculated that if it forms more than one alliance, the government will subsidize the Y3 terminal. Figure 3 presents the area of its core. The intersection point of three segments is its core. Figure 3 shows that the inner area of this triangle expresses how much profit each player derives from the total profit. Consequently, the triangle height is the total profit when all players mutually cooperate. In this case, the height is 855. The height from a side to the inner point signifies the profit a player at an opposite point from the edge receives. For example, the height from parallel line to a point shows k_1's profit.

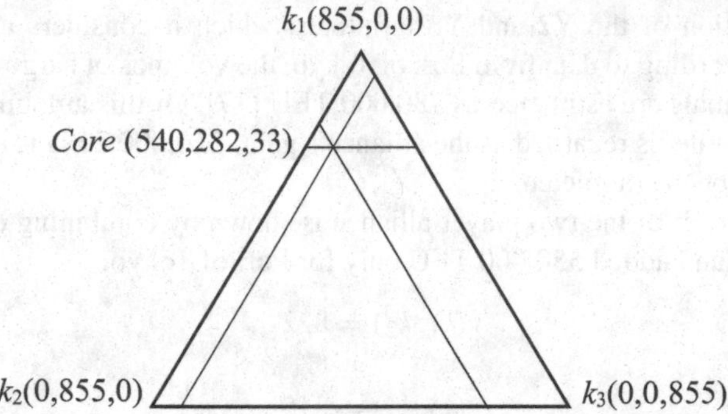

Fig. 3. Barycentric coordinates of player profits.

7 Discussion

In actuality, the alliance itself is far from running smoothly. Particularly in the case of Tokyo Bay, the Port of Tokyo itself has not even joined the alliance. The reason is that the alliance is not attractive to the Port of Tokyo. Even if it did join, it would not be profitable.

As discussed in Sect. 6, the core for the cooperation between ports in Tokyo Bay was found. The core is the set of imputations that are not dominated by other imputations in cooperative game theory, as described in Sect. 3. In other words, no cooperation in the distribution of benefits will occur when cooperation takes place because any imputation other than the core will cause dissatisfaction for a player.

In the original plan, the paper would study the reason why the core was empty or find out which one of the imputations was better by using Shapley value. But in reality, the analysis above could not be conducted since the core is only one imputation and the fact may cause such type of instability. In other words, the stability of the core is easily affected by other factors. The fact that the core is single means that even if the number of members of the alliances increases from two to three, the gain each player does not increase and the distribution of the gain is stable in only one way. As a practical matter, the Port of Tokyo has not joined the alliance of the three ports and there could be several reasons for it.

First, not only players in Tokyo Bay, but MLIT have decided that the purpose of alliance is to semi-nationalize ports using the system, rather than to produce benefits of cooperation. Therefore, considering the merits and benefits of joining the alliance does not occur to them. The YKIP players and the government have not devised such incentives to add the Port of Tokyo to the alliance.

Regarding the second point, in Sect. 6, the gain of each characteristic equation was treated, referring to interview responses and using the same index for each parameter. Freight, subsidies, and rationalization through mergers are described, but others must exist. For example, according to interviews, intervention by the government entails disadvantages. The major shortcoming of government intervention in the Port of Tokyo is that government will have disadvantages in cargo distribution, and the possibility also exists that the government will take control of management. Few examples exist of good relations between the Tokyo Metropolitan Government and the government in all industries, not just ports. This is often the case because Tokyo is better funded than other prefectures. Other costs are also incorporated in coordination between the Port of Tokyo and Rinkai HD. These parameters must be adjusted through more interviews and researches in the future.

Thirdly, this core indicates that these three ports would work together if there were not alliances like YKIP. In fact, interview responses revealed that talks of forming an alliance were progressing before MLIT joined, meaning (whether with profit sharing or not) that not everyone was dissatisfied and that some benefit was to be gained by working together. The reason why these three ports are not cooperating at this stage is that the situation now is that YKIP and Port of Tokyo are not sure whether to cooperate. This calculation resembles that for a two-player game. To cooperate each other, at least, the Port of Yokohama must get some profit by doing so because there are some administrative costs in a coalition. Hence, YKIP side must get some parts of gains of the Port of Tokyo; alternatively, YKIP itself must acquire some advantage. However, results of the interviews showed that YKIP was unable to plan a quantitative benefit. That would be difficult to accomplish in its current state.

8 Conclusions

This study of the current status of Japanese ports particularly addresses Tokyo Bay. Japanese ports lag behind the world standard in terms of cargo handling volume, partly because of delays in responding to the increasing size of container ships and partly because of the failure of the Super Hub Port Policy to foster a Global Terminal Operator (GTO). The analysis examined the International Container Strategy started in 2011 and the designation of Strategic International Ports. Including the ports that have been designated as Strategic International Ports, the research covered ports in three locations: Tokyo Bay, Osaka Bay, and Ise Bay. For convenience in access to develop interviews, the study specifically addressed Tokyo Bay and interviewed officials of the Central Government.

Because the interviews were conducted online, the responses cannot be certified as reflecting the respondent's true opinions. However, respondents supplied many details. Results obtained for the Yokkaichi Port interview a supplementary survey of how the alliance is working in other areas as well as in Tokyo Bay. Conducting additional interviews at other ports was not possible because of COVID19 restrictions.

The results of the interviews clarified that the current alliances at the ports have no cooperative relation for cargo sharing. Regarding YKIP particularly, it turned out that the main purpose of the alliance was to provide subsidies and loans to the Port of Yokohama for construction of deeper water areas and improvement of the Minami-Honmoku Wharf. The Port of Yokohama has abided by the will of the government and is apparently satisfied. The Port of Tokyo, however, was originally willing to join an alliance and exchange cargo with other ports, as in Seattle–Tacoma and New York – New Jersey. For NYP, a common understanding exists because it was a recipient of such subsidies.

Based on interview results and the estimation of respective benefits, the alliance in Tokyo Bay was estimated by analyzing it from a cooperative game theory approach. In addition to the volume of cargo handled, which is often used to evaluate ports, many other parameters such as subsidies for Strategic International Ports, interest-free loans, cargo evaluation were used. Those parameters reflect the area behind the port, management initiatives, and the benefits of forming an alliance with a large port (for a small port).

It is questionable whether a model can be created that perfectly matches reality. However, once the relation between the size of the gain and the difference between the case of forming an alliance and the case of not forming an alliance are clarified, it is expected to be possible to adjust the model through negotiations in real situations.

In addition, the government was not included as a player in this analysis based on cooperative game theory. In the first place, however, the International Container Strategic Port Policy is a policy initiated by the central government. Its purpose is to maintain trunk routes and thereby maintain and develop Japanese industries. Because governments have invested great amounts of funds in the project, one can calculate the project benefits in terms of the development of Japanese industries in the cooperative game.

For example, for Tokyo Bay, by adding another player, the government, it is expected to be possible to examine what incentives should be awarded by the government to achieve cooperation among the three ports. Particularly, incentives must be offered that would outweigh the negative effects on forces that view entry itself as negative. In any case, this study has only calculated gains in terms of the volume of containers handled. If the government is to be included as a player, then the issue will be how to incorporate factors such as trunk routes, industrial development, and initiatives.

Finally, because of COVID-19, additional planned interviews were restricted or canceled. For that reason, the interviews and information to support these analyses might be few. Initially, all field visits were planned, but as described above, because of pandemic restrictions, the research was conducted online. For future studies, interviews must be extended to include other ports and to incorporate the views and experiences of other factors and players.

References

1. Stopford, M.: Maritime Economics, 3rd edn. Routledge, New York (2008)
2. Inoue, S.: Realities and challenges of port alliance in Japan -ports of Kobe and Osaka. Res. Transp. Bus. Manag. **26**, 45–55 (2018)
3. Hatani, F.: Institutional plasticity in public-private interactions: why Japan's port reform failed. J. World Bus. **51**(6), 923–936 (2016)
4. Takahashi, H., Akakura, Y.: Analysis of trends in the increase in the size of containerships calling at Japanese ports in Japanese. Transp. Policy Res. **2–4**, 30–39 (1999)
5. Kondo, T.: History of Port Policy and Future Challenges - Prospects for Strategic International Container Ports in Japanese (2010). https://www.sangiin.go.jp/japanese/annai/chousa/rippou_chousa/backnumber/2010pdf/20101101041.pdf. Accessed 5 June 2021
6. Brooks, M.R., Cuillinane, K.P., Pallis, A.A.: Revisiting port governance and port reform: a multi-country examination. Res. Transp. Bus. Manag. **22**, 1–10 (2017)

7. M. o. L. I. T. a. T. Port and Harbor Bureau: Transportation Policy Review Meeting 49th meeting of the Port and Harbor Division Reference 4–1 About the Port Management Company System in Japanese, 5 July 2014. https://www.mlit.go.jp/common/000185169.pdf. Accessed 5 June 2021
8. Geospatial Information Authority of Japan. https://maps.gsi.go.jp/#5/36.049099/136.845703/&base=std&ls=std&disp=1&vs=c1j0h0k0l0u0t0z0r0s0m0f0. Accessed 5 June 2021
9. Watanabe, T., Kawasaki, T., Matsuda, T.: An analysis of port management methods focusing on multi-actor relationships in integrated ports in Japanese. J. Log. Ship. Econ. **53**, 31–40 (2019)
10. Park, N.K., Suh, S.C.: Port competition study: cooperative game model. J. Contemp. Manage. **4**(3), 38–52 (2015)
11. Song, D.-W., Panayides, P.M.: A conceptual application of cooperative game theory to liner shipping startegic allianaces. Maritime Policy Manage. **29**(3), 285–301 (2002)
12. Morgenstern, O., Neumann, V.J.: Theory of games and economic behavior (1953)
13. Nagarajan, M., Sošić, G.: Game-theoretic analysis of cooperation among supply chain agents: review and extensions. Eur. J. Oper. Res. **187**, 719–745 (2008)
14. Cachon, G.P., Nettessine, S.: Game theory in supply chain analysis. Inf. Tutor. Oper. Res., 200–233 (2014)
15. Mikio Nakayama, Y.H.M.: Cooperative Game Theory in Japanese, Keiso Shobo (2008)
16. Ministry of Land, Infrastructure, Transport and Tourism: 58th Harbour Subcommittee; Revision of the Port Plan for Keihin Port, Strategic International Port 2–1 in Japanese, 14 November 2014. https://www.mlit.go.jp/policy/shingikai/port01_sg_000160.html. Accessed 5 June 2021
17. Kanto Regional Development Bureau: Kanto Regional Development Bureau Project Evaluation Monitoring Committee; Port of Tokyo Central Breakwater Outer Area International Marine Container Terminal Development Project in Japanese, 27 November 2014. https://www.ktr.mlit.go.jp/ktr_content/content/000113588.pdf. Accessed 5 June 2021
18. Ishii, M., Tezuka, P.T.-W.L., Chang, Y.-T.: A game theoretical analysis of port competition, vol. 49, no. 1, pp. 92–106 (2013)
19. Cui, H., Notteboom, T.: A game theoretical approach to the effects of port objective orientation and service differentiation on port authorities' willingness to cooperate. Res. Transp. Bus. Manage. **26**, 76–86 (2018)
20. Bureau of Port and Harbor of Tokyo: Port Statistics of the Port of Tokyo/Container handling status (2019). (in Japanese). https://www.kouwan.metro.tokyo.lg.jp/yakuwari/toukei/. Accessed 5 June 2021
21. Bureau of Port and Harbour of Kawasaki: The port of Kasaki's statistics/3.Container in Japanese (2019). https://www.city.kawasaki.jp/580/page/0000122979.html. Accessed 5 June 2021
22. Bureau of Port and Harbour of Yokohama: Latest statics/Container in Japanese (2021). https://www.city.yokohama.lg.jp/city-info/yokohamashi/yokohamako/kkihon/tokei/statistics.html. Accessed 5 June 2021

Improving the Elicitation Process for Intra-criterion Evaluation in the FITradeoff Method

Paolla Polla Pontes do Espírito Santo[1,2](✉), Eduarda Asfora Frej[1,2], and Adiel Teixeira de Almeida[1,2]

[1] Departamento de Engenharia de Produção, Universidade Federal de Pernambuco, Av. Da Arquitetura-Cidade Universitária, Recife, PE, Brazil
paolla.polla@ufpe.br, {eafrej,almeida}@cdsid.org.br

[2] Center of Decision Systems and Information Development-CDSID, Universidade Federal de Pernambuco, Av. Da Arquitetura-Cidade Universirtária, Recife, PE, Brazil

Abstract. With the advance coming from studies in the area of decision making, different models have emerged to assist in the interpretation of multicriteria decision problems. One of the most recent improvements in the MCDM/A mathematical models deals with the use of partial information about the preferences of the decision makers at the elicitation process. The FITradeoff method is a MAVT (Multi-attribute Value Theory) method that requires only strict preferences and uses partial information in judgments, reducing the amount of information required. Therefore, this study aims to improve the intra-criteria evaluation step of the FITradeoff method, by proposing a new approach for elicitation of marginal value functions based on partial information. The proposed approach is based on the traditional bisection method, but requires preference statements only. The results obtain show that the approach using the bisection method associated with the use of partial information appears to have a good performance, enabling the improvement of the process in terms of reducing the effort and time required.

Keywords: Bisection method · Intra-criteria evaluation · Partial information · FITradeoff method

1 Introduction

Decision making is an essential cognitive process of human beings (Zuheros et al. 2020). With the advance coming from studies in this area, different models have emerged to assist in the interpretation of multicriteria decision problems. Analyzing the widest possible range of alternatives and solving them according to multiple criteria of interest, generally conflicting, with one or more decision-makers.

Thus, there is a variety of elicitation procedures that use different tools to obtain the expectations and necessities of its users. de Almeida, Geiger and Morais (2018) show that one of the most recent improvements in the MCDM/A mathematical models deals

A. T. de Almeida and D. C. Morais (Eds.): INSID 2021, LNBIP 435, pp. 68–86, 2021.
https://doi.org/10.1007/978-3-030-91768-5_5

with the use of partial information about the preferences of the decision makers (DM) at the moment of elicitation.

Regarding the additive multicriteria methods, after the problem has been well structured, the first step to start eliciting preferences is the intra-criterion evaluation (de Almeida et al. 2021). They can be performed in different ways, for example, with the construction of qualitative scales, through indirect evaluations-Bisection Method and Differences Method, or by direct evaluation (Belton and Stewart 2002).

However, in an attempt to simplify the elicitation procedure, several additive MCDM methods reduce this assessment considering only the linear form of the value function for the criteria, obtained based on a normalization process. This simplification introduces modeling errors but reduces elicitation errors (de Almeida et al. 2021). Toubia et al. (2013), highlight that these simplifications can limit the performance of the analyzes realized.

Several decision support methods have been developed to aid the DM in solving multicriteria problems, offering them structured approaches. One of these is the Flexible and Interactive Tradeoff elicitation- FITradeoff (de Almeida et al. 2016; Frej et al. 2019). The FITradeoff is a MAVT method that requires only strict preferences statements and uses partial information in judgments, reducing the amount of information required (Pergher et al. 2020). Consequently, demanding less cognitive effort from the decision maker, leading to fewer inconsistencies during the elicitation process.

Therefore, the present study aims to improve the intra-criteria evaluation step of the FITradeoff method, with a flexible elicitation procedure that uses the bisection method with partial information to construct non-linear value functions.

This article is structured as follows. Section 2 presents the FITradeoff method, commenting on some studies with real applications that used it. Section 3 shows the intra-criterion evaluation in additive models, presenting it in the context of partial information and with the bisection method. Section 4 describes a new approach for intra-criterion evaluation with partial information, followed by a numerical application, in Sect. 5. Finally, Sect. 6 presents the final comments and highlights future research.

2 Flexible and Interactive Tradeoff

The Flexible and Interactive Tradeoff method (de Almeida et al. 2016) is based on the classic tradeoff procedure (Keeney and Raiffa 1976), being a MAVT method with considers that decision makers have compensatory rationality, i.e., they admit that low performance in one criterion can be compensated by high one in another, and presents an additive aggregation model (Pergher et al. 2020). The FITradeoff solves MCDM/A problems with partial information from the DMs (de Almeida et al. 2016). Assuming an MCDM/A problem with m alternatives and n criteria, the MAVT procedure is illustrated in Eq. (1), where a_j is an alternative to the set of m alternatives, k_i is the scale constant of criterion i, and $v_i(x_{ij})$ is the value of consequence of alternative j in criterion i, normalized in an interval 0–1 scale, defined according to a marginal value function. Thus, the best alternative of the set is the one with the highest global value $V(a_j)$ (Roselli and de Almeida 2021).

$$V(a_j) = \sum\nolimits_{i=1}^{n} k_i v_i(x_{ij}) \quad (1)$$

The main difference in relation to previous studies is related to the elicitation process. de Almeida et al. (2016) present the concept of flexible elicitation, a constructive context of multicriteria value models which allows the consequences between alternatives to be compared, exploring strict preferences statements instead of indifference.

In this way, possibilities are considered such as that of the decision maker not being familiar with certain methods or cannot provide information and visualize more satisfactory and real results when less cognitively demanded. Frej, Ekel and de Almeida (2021) argue that the development of an approach to deal with partial information is a constructive way to apply traditional MCDAs with elicitation techniques that significantly reduce the time and efforts required.

The use of partial information is based on preference relations to find a solution, which in most cases can be achieved by incomplete information declared from the decision maker. And these are used to solve a linear programming problem (LPP) (de Almeida et al. 2016). The problems can be classified as of choice (de Almeida et al. 2016), ranking (Frej et al. 2019) or sorting (Kang et al. 2020).

Frej et al. (2019) explain that the LPP referring to the choice problems aims to use the concept of potential optimality, finding at the ending of the procedure an optimal solution or set of potentially optimal alternatives. The ranking problematic uses the concept of pairwise dominance relations to find a complete or partial (pre)order ranking of alternatives. While for the sorting problems, Kang et al. (2020) present the use of border values that limit the consecutive classes of problems.

The FITradeoff method has been used to solve several multicriteria problems in different areas of expertise. For example, an application for supplier selection (Frej et al. 2017), in the selection of programming rules (Pergher et al. 2020), applications in the textile sector (Rodrigues et al. 2020), real cases in the energy sector (Fossile et al. 2020), system design studies using neuroscience experiments (Roselli et al. 2019a,b) and prioritizing Brazilian Federal Police operations (Cunha et al. 2020).

The method is embedded in a Decision Support System (DSS), which is available at www.cdsid.org.br/fitradeoff. The DSS uses the concept of flexible elicitation. The flexibility in this consists in systematically assessing the possibility of finding a solution to the problem during the elicitation process. The procedure can be interrupted as soon as a solution is found or until the moment when the DM wants to provide information (de Almeida et al. 2016). During the elicitation process, partial results can be viewed using tables and graphs. Displaying the information processed in different ways, helping the decision maker to understand the performance of the alternatives about each evaluated criterion (Roselli et al. 2019a,b).

Regarding the intra-criterion evaluation stage, the FITradeoff method was originally conceived to allow the incorporation of non-linear value functions, since the whole structure of the classical tradeoff procedure is preserved. The current version of the FITradeoff DSS, however, consider the incorporation of non-linearity in the value function throughout a direct specification of the form of the function by the DM, which can be of four different types: linear, exponential, logarithmic, and logistic. When non-linear functions are declared, the decision maker is asked to assign values of parameters. However, these values may not be precisely known, or the DM may not be willing to provide them. Therefore, there is an opportunity to improve the intra-criteria evaluation process of FITradeoff, by allowing the DM to elicit marginal value functions, instead of direct specifying them. However, this elicitation process should be carried out considering partial information, as well as the intercriteria evaluation does, in order to keep the basic premises of the method of saving time and effort from DMs.

Based on this motivation, the study proposes a new approach to improve the intra-criterion evaluation in FITradeoff, based on the well-known bisection method (Belton and Stewart 2002), but considering partial information from the DMs. Hence, the proposed approach works with preference statements obtained from the DM, instead of indifferences points required by the classical bisection method. It is intended that, from the beginning of elicitation, the values and forms of the functions of each criterion more faithful reflecting the relations between the decision maker preferences and the final model of his problem. Furthermore, making use of partial information reduces the amount of direct information required from the DM, consequently reducing the cognitive effort required during the procedure.

3 Intra-criterion Evaluation in Additive Models

The intra-criterion evaluation of additive model for aggregation of criteria consists of establishing the value function of each criterion, including cases where this function is non-linear. The value function methods synthesize the evaluation of the performance of the alternative against individual criteria, together with inter-criterion information, providing an overall evaluation of each alternative indicated of the decision makers' preferences. Once the scale reference points are determined, it should be considered how the other scores will be assessed. It can be done in three ways: (a) definition of a partial value function, (b) construction of a qualitative value scale, or (c) direct evaluation of alternatives (Belton and Stewart 2002).

The first step to defining a value function is identifying a measurable attribute scale that is closely related to the decision maker values. The partial value function reflects the preferences of decision makers at different levels of aspiration on the measurable scale. It can be evaluated directly or through indirect evaluation. Direct assessment usually uses a visual representation. About indirect evaluation, the bisection method is one of the widely used methods (Belton and Stewart 2002).

In the bisection method, the decision maker is asked to define a point on the attribute scale that is halfway in terms of value between the two endpoints, obtaining two linear partial value functions. This process can be repeated several times until the decision maker is indifferent between the partitions (Groothuis-Oudshoorn et al. 2017). Belton and Stewart (2002) also state that usually with five points it is possible to provide enough information to the analyst to find the value functions.

This is generally used in elicitation procedures that enable linearized and non-linear functions and permit the search for behaviors that more accurately reflect the preferences of the decision maker. Thus, using this method to identify the behavior of the partial value function of criteria, in the intra-criteria evaluation stage of multicriteria problems, may prove to be especially suitable.

However, its main disadvantage is the requirement of indifference points when comparing the performances between alternatives, generating inconsistencies during the elicitation process, because it requires major cognitive effort on the decision maker when requiring complete information (de Almeida et al. 2016; Roselli et al. 2019a,b). Thus, the development of an intra-criteria elicitation procedure that reconciles the application of the bisection method to the use of partial information can be relevant, as it would allow the decision maker to declare their aspirations and behaviors from the intra-criterion evaluation stage.

3.1 Partial Information in the Intra-criterion Evaluation Stage

In the intra-criteria evaluation, some methods consider a simplified approach when assuming linear value functions, as in the SMARTS and SMARTER methods (Edwards and Barron 1994). Another group of methods builds the value function based on pairwise comparisons between preference statements, such as the analytical hierarchy process AHP and MACBETH (Vasconcelos and Mota 2019). Outranking problems or multiobjective mathematical programming, seek to identify upper, lower, and/or veto thresholds that reflect the interests of a decision maker, for these attribute values.

In the literature, it is possible to identify the increasing use of partial information, due to the use of strict preference statements during an interactive process between the decision maker and analyst. Making the procedure less stressful and less susceptible to inconsistencies. The use of partial information in the inter-criterion evaluation stage is widespread in the literature when determining the ordering of a problem's criteria and their respective values. When contextualized in the intra-criterion evaluation stage, it is noted that studies have been exploring this potential better.

Jaszkiewicz and Slowinski (1997) presented an interactive procedure, the LBS-Discrete, for the analysis of a multicriteria agricultural problem. The procedure is an extension of multiobjective linear programming (PLMO) Light Beam Search, being non-linear for the discrete case. To ensure an easily assessment for the decision maker, the authors considered preference statements at the steps intra and inter-criterion information for the set of points analyzed in the sample, updating the space of solution for each question asked. In the rounds, the decision maker determined the upper and lower bounds of the permissible solution space. The procedure could be interrupted if the DM wished.

Eum et al. (2001) provided an extended outranking model to establish the potential optimization of alternatives in the analysis of the multicriteria decision. Assuming that in problems with partial information, not only are the weights of attributes are imprecise known, but also their marginal values. In this way, the resulting model became a non-linear programming problem being transformed to an equivalent LPP. To demonstrate the method, the authors solved problems found in the literature.

Lahdelma et al. (2003) describe the SMAA-O method. Designed for problems where weights are not precisely known and criteria information is partially or integrally ordinal, making the DM to list alternatives in terms of ratings for some or all criteria. To modeling the value function of these criteria, numerical mappings were created that generated stochastic cardinal values corresponding to the ordinal values. In the end, a problem of the selection of a solid waste management system was applied.

Narula et al. (2004) developed an interactive learning-oriented method for solving MCDA problems with many alternatives and few criteria. Where it is possible for the DM to successively evaluate small sets of alternatives, systematically, specifying only the information that wishes or changes considered acceptable for the values and direction of the criteria involved. With the aid of software, at each iteration the decision maker compared neighboring groups of alternatives, ordering them, solving a scalarization of the problem.

Thus, it is considered relevant that the context of partial information is also explored in the stage of intra-criteria assessment of MAVT problems, making the elicitation stage more realistically for the preferences of the decision makers. By demanding them less cognitively, so exploring gradually the space of action of their problems, becoming a learning process.

3.2 Bisection Method in Interactive Procedures

In recent decades, research has shown the desire to understand in a more real way how decision makers behave in the face of not fully understanding aspects of their multicriteria problems, making use of partial information and flexible elicitation procedures, in different decision methods and methodologies.

Approaches based on problems with dynamic systems and Utility Theory have also been exploring solutions that consider issues that are normally dealt with a deterministic vision in a more realistic way. Some of the resources explored to structure these problems include the use of analytical and/or statistical tools, the bisection method - traditional or improved - and inferences without parametric equations.

Toubia et al. (2013) propose a dynamic methodology to relate time and risk parameters in decision making. The use of pre-computed tables of possible preference questions to a decision maker is implemented, as the latter provides answers. Designing such choices to optimize the information provided, while taking advantage of the distribution of parameters, capturing the deviations between responses.

Chapman et al. (2018) present the DOSE-Dynamically Optimized Sequential Experiment- estimating the preference parameters accurately and quickly when selecting a personalized sequence of simple questions for each participant. The method used a parametric structure and Bayesian computation, to dynamically select a sequence from a set of statements. The process is interactive, updating the problem's constraints space until a predetermined number of questions or when the parameters are found.

Recently, Bertani et al. (2020) identified values and behavior of the weighting function, parameterizing it through a family of linear splines that can return smooth nonlinear shapes. Thus, the permissible limits were obtained as the solution to problems of restricted linear optimization. The judgments of decision makers were captured using the bisection method with partial information, to identify the space of actions of the problem. Some questions of preference were defined a priori.

Oliveira and Dias (2020) found consumer preferences for alternative fuel vehicles through a MAUT-based approach. The authors use the bisection method to obtain utility and tradeoff functions for calculating the scale constants of the attributes. Belton and Stewart (2002) considered that one of the possible areas of research in the MCDA area would be the identification of general weaknesses in decision support models. Groothuis-Oudshoorn et al. (2017) point out that a structural source of problems, in the performance evaluation stage, in the form of the value function, as it is normally assumed to be linear.

4 A New Approach for Intra-criterion Evaluation with Partial Information

Since the value function should represent the preferences of a decision maker measurably, in terms of aspiration, the bisection method is applied to determine points on the scale considered, outlining the partial value functions. And finally, to identify the behavior described by a criterion when it is elicited.

The proposed approach follows a dynamic similar to that found in the literature for the traditional bisection method (Belton and Stewart 2002; Groothuis-Oudshoorn et al. 2017; Bertani et al. 2020), however considering partial information from the DM. Two reference values are compared and the decision maker is asked why there is a greater predilection. However, instead of a point of indifference, want to find ranges of values through strict preference statements.

So, initially, the question has the basic structure: "What do you prefer, increase the value of the consequence in the Ci criterion from A to X or from X to B?". For minimization criteria, the term increase is replaced by decrease/minimize. Concerning the reference value X, it is updated to reduce the interval between the lower and upper bounds obtained with each answer given. For illustrative purposes, updates based on the answers given by the decision maker are made using the following logic:

Question 1: "What do you prefer, increase the value of the consequence in the Ci criterion from A to $X_n$or from $X_n$to B? DM: I prefer to increase from A to X_n". Range1 → A to X_n.

Question 2: "What do you prefer, increase the value of the consequence in the Ci criterion from A to $X_{n/2}$or from $X_{n/2}$to B? DM: I prefer to increase from $X_{n/2}$to B". Range 2 → $X_{n/2}$to X_n.

In this case, there was an update of the lower bound, because when answering, the decision maker migrated his preference interval to the upper segment of reference X. Similarly, the upper bound is updated when the chosen interval returns to the lower segment of the reference. And so, successively, until the stopping criterion is met or the decision maker does not wish to proceed. This procedure is performed until the last point is inferred.

About the number of points, the literature usually considers that five points provide sufficient information for the shape of a value function to be identified (Belton and Stewart 2002). In this proposal for the bisection method with partial information, the first and last points of the scale (0–1) will be determined at the local scale. Thus, the worst and best values of the consequences reported in the problem will be adopted as X = 0 and X = 1, depending on the direction of the criterion (minimization or maximization). Remaining the elicitation of points X = 0.25, X = 0.5 and X = 0.75.

4.1 Intra-criterion Elicitation Procedure

After declaring the values of the consequence matrix, the DM provides information for three rounds j, each with two stages, to identifying three points, in addition to the local extremes of the scale. However, the elicitation not be conducted to determine indifference statements, but an admissible range of values for the DM, decreasing his cognitive effort by request only strict preference statements.

The stopping criterion to change between the rounds can occur in two ways: i) assuming a percentage margin (P) of 5%, 10%, 15%, or 20% over the range R, between the maximum and minimum limits of the consequence values for a criterion Ci. The decision maker should define it before starting the elicitation; or ii) Anytime the decision maker wants to stop answering the intra-criteria elicitation questions, as the procedure supports partial information and is flexible. In this way, the margin admitted for variation is defined as a stopping criterion in Step1 of all rounds of the intra-criterion elicitation procedure. Initially, the interval analyzed for asking the questions will vary from A to B.

For each answer given, the lower and upper limits of the interval are checked and updated, when possible, i.e., each question is generated to decrease the numerical value between the lower limit and the upper limit of the range generated with the bisection method using partial information. Until a value equal to or less than the stopping criterion is reached. If the DM has not interrupted the process and the value is to be true, Step 2 of the procedure begins.

This step consists of presenting a graph with three shapes that describe possible behaviors of the criterion under analysis so that the decision maker chooses the one that he/she judges closest to your preferences. The series of round 1, for the point $X_{0.5}$, are built with as illustrated in Fig. 1.

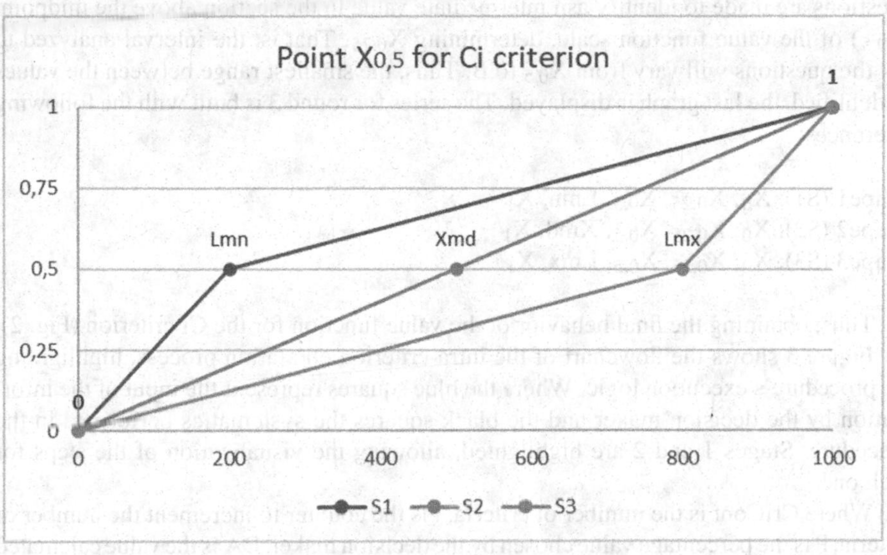

Fig. 1. Graph example with series of point $X_{0.5}$

Where the value 0 is the lower bound of the local scale of consequences. It is the worst consequence value declared in the matrix; the value 1 is the upper bound of the local scale of consequences. It is the best consequence value declared in the matrix. And, obtained in Step 1 of the round, LMn: Minimum range limit, LMx: Maximum range limit, and Xmd: Midpoint of the range.

Once the series is selected, the point $X_{0,5}$ will assume the value of LMn or LMx or Xmd, depending on the choice made, starting round 2. In this step, the elicitation process of Step 1 occurs similarly to that described for round 1, however, questions are asked to identify an intermediate value in the section below the midpoint ($X_{0.5}$) of the value function scale. Identifying the reference of $X_{0.25}$. That, the analyzed interval to ask the questions will vary from A to $X_{0.5}$.

For each answer given, the lower and upper limits are checked and updated, when possible. Until a value equal to or less than the stopping criterion is reached or the decision maker interrupts the process. Thus, Stage 2 of round 2 is initiated and again a graph is presented so that the DM chooses the best. The series for round 2 is built with the following references:

Shape1 (S1): X_0, Lmn, $X_{0,5}$, X_1
Shape2 (S2): X_0, Xmd, $X_{0,5}$, X_1
Shape3 (S3): X_0, Lmx, $X_{0,5}$, X_1

Where $X_{0,5}$ is the value chosen in round 1, being the midpoint in terms of local scale. The other parameters remain with the same interpretation. Once the series is selected, point $X_{0,25}$ will assume the value of LMn or LMx or Xmd, depending on the choice made, starting round 3. Finally, the last point of Step 1 in the process is elicited, but now the questions are made to identify asn intermediate value in the section above the midpoint ($X_{0.5}$) of the value function scale, determining $X_{0.75}$. That is, the interval analyzed to ask the questions will vary from $X_{0.5}$ to B. Thus, the smallest range between the values is identified, the last graph is displayed. The series for round 3 is built with the following references:

Shape1 (S1): X_0, $X_{0.25}$, $X_{0,5}$, Lmn, X_1
Shape2 (S2): X_0, $X_{0.25}$, $X_{0,5}$, Xmd, X_1
Shape3 (S3): X_0, $X_{0.25}$, $X_{0,5}$, Lmx, X_1

Thus, obtaining the final behavior of the value function for the Ci criterion (Fig. 2).

Figure 3 shows the flowchart of the intra-criterion elicitation process, highlighting the procedure's execution logic. Where the blue squares represent the input of the information by the decision maker and the black squares the systematics performed in the procedure. Stages 1 and 2 are highlighted, allowing the visualization of the steps for each one.

Where CriCont is the number of criteria, i is the counter to increment the number of criteria, P is the percentage value chosen by the decision maker, DA is the value calculated to be the stopping criterion, Q is the counter to increase the number of questions, n is the number of rounds, R is the range between high and low bounds. And J is the counter to increase the number of rounds, where, X1 equivalent to $X_{0.5}$; X2 to $X_{0.25}$, and X3 corresponds to the $X_{0.75}$.

After the decision maker inputs the matrix of consequences for the problem, Stage 1 of the intra-criterion elicitation procedure is started. Initially, the DM will define the

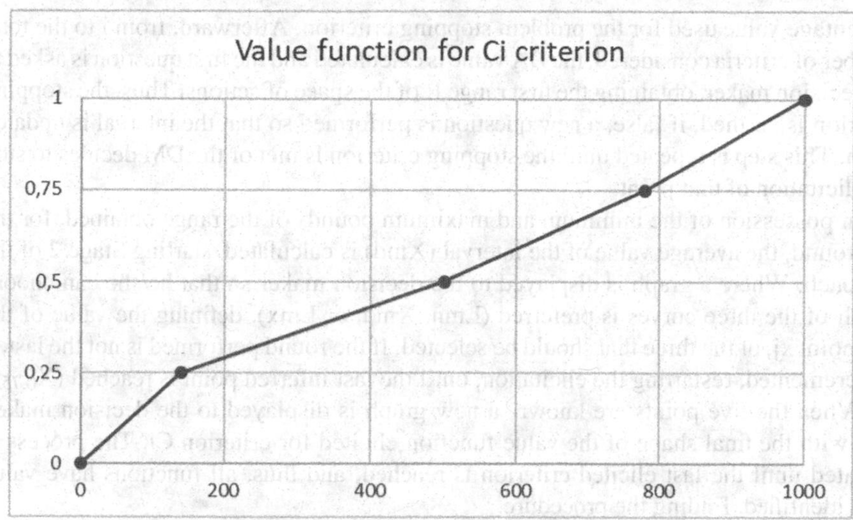

Fig. 2. Final graph example with Ci criterion behavior.

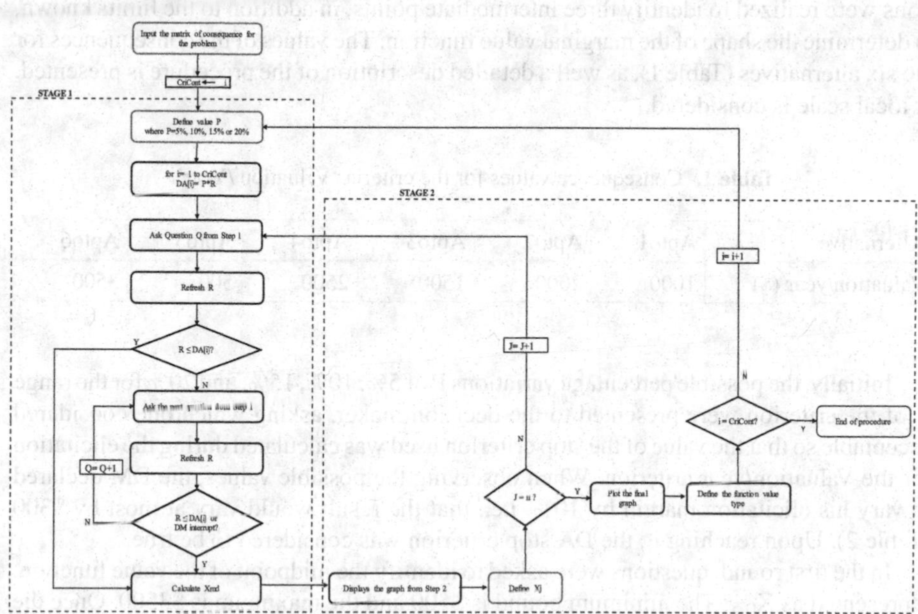

Fig. 3. Flowchart of procedure for intra-criterion evaluation.

percentage value used for the problem stopping criterion. Afterward, from i to the total number of criteria considered, the DA value is calculated and the first question is asked to the decision maker, obtaining the first range R of the space of actions. Thus, the stopping criterion is verified, if false, a new question is performed so that the interval is updated again. This step is repeated until the stopping criterion is met or the DM decides to stop the elicitation of that point.

In possession of the minimum and maximum bounds of the range obtained, for the first round, the average value of the interval (Xmd) is calculated, starting Stage 2 of the approach. Where a graph is displayed to the decision maker so that he/she can choose which of the three curves is preferred (Lmn, Xmd, or Lmx), defining the value of the first point xj, of the three that should be selected. If the round performed is not the last, it is incremented, restarting the elicitation, until the last inferred point is reached ($X_{0.75}$).

When the five points are known, a new graph is displayed to the decision maker, now with the final shape of the value function elicited for criterion Ci. The process is repeated until the last elicited criterion is reached, and thus, all functions have value been identified. Ending the procedure.

5 Numerical Example

In order to illustrate the applicability of the proposed approach, let us consider a multicriteria problem when deciding on renting an apartment. The process of eliciting the continuous maximization criterion Valuation/year is illustrated. Three rounds of questions were realized to identify three intermediate points, in addition to the limits known, to determine the shape of the marginal value function. The values of the consequences for the six alternatives (Table 1), as well a detailed description of the procedure is presented. A local scale is considered.

Table 1. Consequence values for the criteria 'Valuation /Year'.

Alternative	Apto1	Apto2	Apto3	Apto4	Apto5	Apto6
Valuation/year ($)	1000	2000	1500	2500	500	3500

Initially, the possible percentage variations P of 5%, 10%, 15%, and 20% for the range R of the criterion were presented to the decision maker, asking which one considered acceptable so that the value of the stop criterion used was calculated during the elicitation for the Valuation/year criterion. When observing the possible values, the DM declared to vary his elicitation margin by 10%, i.e., that the result would vary at most by $300 (Table 2). Upon reaching it, the DA stop criterion was considered to be true.

In the first round, questions were asked to identify the midpoint of the value function, represented as $X_{0.5}$. The minimum bound is $500 and the maximum is $3500. Once the procedure was initiated, the first question asked in Stage 1 was "*What do you prefer, increase the value from $500 to $2000 or from $2000 to $3500?*" The decision maker declared that he preferred the increase from $500 to $2000. Determining the first range (I1) from $500 to $2000.

Table 2. Calculation of the DA value

P Value (%)	DA Value ($)
10	(3500–500) * 0,1 = 300

Then was asked, *"What do you prefer, increase the value from $500 to $1500 or from $1500 to $3500?"* The decision maker answer that he preferred the increase from $1500 to $3500. Thus, as the response migrated the interval to the upper section of the midpoint of the range, the lower bound of I1 was updated. And the new range I2 being $1500 to $2000.

The third question asked was *"What do you prefer, increase the value from $500 to $1600 or from $1600 to $3500?"*. The DM declared that he preferred the increase from $1600 to $3500. Thus, again the lower limit of the admissible space has been updated and the new range I3 ranging from $1600 to $2000, respectively, the lower and upper bounds.

In the fourth question, it was asked *"What do you prefer, increase the value from $500 to $1900 or from $1900 to $3500?"* The decision maker replied that he preferred the increase from $500 to $1900. Thus, the interval returned to the lower section of the midpoint of the range, updating the upper bound. And the range I4 staying $1600 to $1900. At the fourth question, it was verified that the DA value for interval I4 was true. Finishing Step 1 of the elicitation for point $X_{0.5}$, calculating the Xmd value.

In this way, the stage 2 of round 1 was started, where the graph with the plot of the three points known in I4 (Lmn, Lmx, Xmd) was displayed to the decision maker (Fig. 4) so that he could choose the best shape, setting the value to $X_{0.5}$.

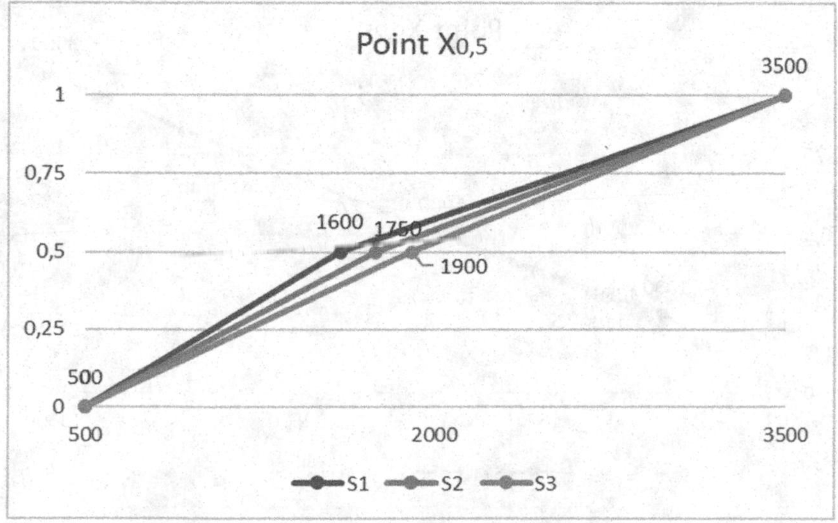

Fig. 4. Graph with series of point $X_{0.5}$

Observing the graph, the DM opted for the behavior expressed with the value of \$1900 to $X_{0.5}$. Justifying being the one with the lowest convexity. Then, round 2 of the procedure was initiated, determining the point $X_{0.25}$.

For the second round, in Step 1, questions were asked to identify the intermediate value in the section below the midpoint $X_{0.5}$. The range considered for asking the questions ranged from \$500 to \$1900, respectively, the lower and upper bounds observed. Where \$1900 was taken from the I4 range, round 1.

The first question asked was "*What do you prefer, increase the value from \$500 to \$1200 or from \$1200 to \$1900?*" The decision maker answer that he preferred the increase from \$500 to \$1200. Thus, the interval I1 was defined between \$500 and \$1200, respectively, with the lower and upper limits of the first interval. Then was asked, "*What do you prefer, increase the value from \$500 to \$1100 or from \$1100 to \$1900?*" The DM declared that he preferred the \$500 to \$1100 increase. In this way, the upper limit of the R range has been updated and the value obtained for the new range I2 from \$500 to \$1100.

Finally, the last question was "*What do you prefer, increase the value from \$500 to \$1000 or from \$1000 to \$1900?*" The decision maker stated that he preferred the increase from \$1000 to \$1900. Thus, the answer was moved to the upper section of the reference and the lower bound considered was updated. In the end, the range I3 was \$1000 to \$1100.

In this round, with one less question in relation to round 1, it was verified that the DA value for the interval I3 was reached, being below the stopping criterion definite. At the end of Step 1 of the elicitation for point $x_{0.25}$, the value of Xmd was calculated. Thus, stage 2 of round 2 was initiated, where the graph with the plot of the three points known in I3 (Lmn, Lmx, Xmd) was displayed to the DM (Fig. 5).

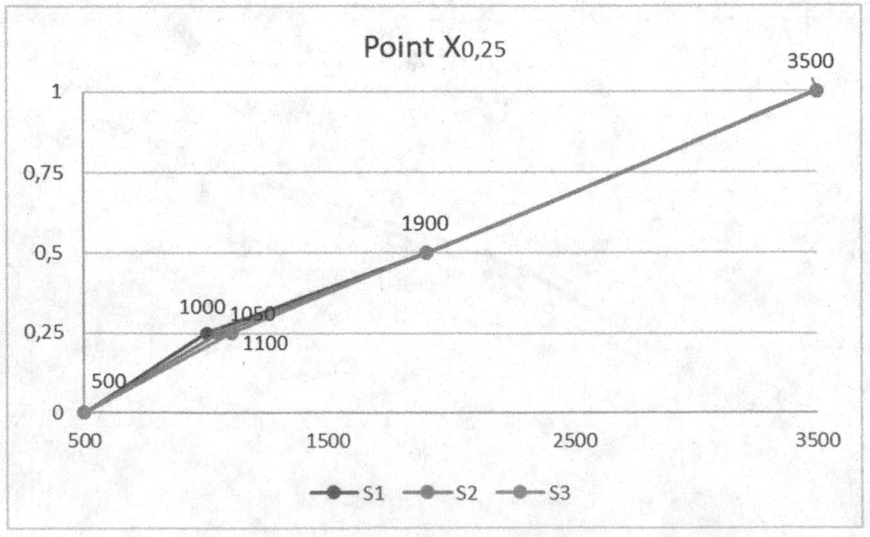

Fig. 5. Graph with series of point $X_{0.25}$

As noted, the elicitation of this point reached a very small variation between the limits of the interval. And when analyzing the graph, the decision maker opted for the behavior expressed with the value of $1100 to $X_{0.25}$. Again, choosing the curve with the smallest convexity. Then, the last round is described, identifying the point $x_{0.75}$.

Finally, in the third round, questions were asked to identify an intermediate value in the section above the midpoint of the value function scale. Thus, the interval considered for performing the questions in Step 1 ranged from $1900 to $3500, the upper and lower bounds, respectively.

Initially, the decision maker was asked, *"What do you prefer, increase the value from $1900 to $2700 or from $2700 to $3500?"* The DM declared that he preferred the increase from $1900 to $2700, respectively, the lower and upper bounds of the first range I1. The second and final question was *"What do you prefer, increase the value from $1900 to $2600 or from $2600 to $3500?"* The decision maker said preferred the increase from $2600 to $3500. Thus, with the answer given, the interval changed to the lower section of the reference, updating the lower limit of the space considered. In the end, the range I2 for point $x_{0.75}$ was between $2600 and $2700.

With two questions, the DA value was reached in round 3, presenting a range of only $100.00, i.e., 2/3 below the value determined by the decision maker. In this way, Step 1 of elicitation for point $X_{0.75}$ was completed and the Xmd was calculated. Starting stage 2 of round 3, where the graph with the plot of the three points known in I2 (Lmn, Lmx, Xmd) was displayed to the decision maker (Fig. 6).

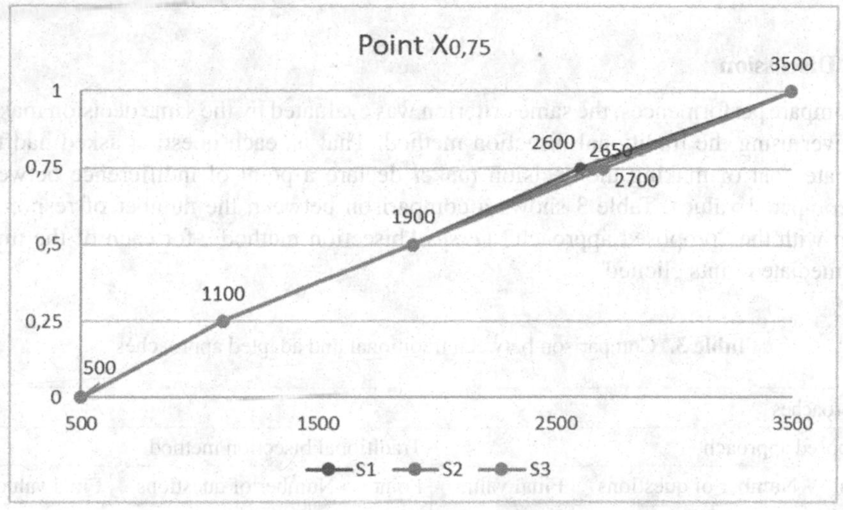

Fig. 6. Graph with series of point $X_{0,75}$.

As in the previous round, the elicitation allowed a small gap between the limits of I2. This being one of the reasons why the decision maker chose the behavior of the curve with the value of $X_{0.75} = \$2.650$, the midpoint. After all rounds and stages were completed, the final graph (Fig. 7) in the form of the value function for the maximization criterion

'Valuation/year' was displayed to the decision maker, presenting that the function can correspond to a logarithmic behavior. Ending the intra-criterion elicitation.

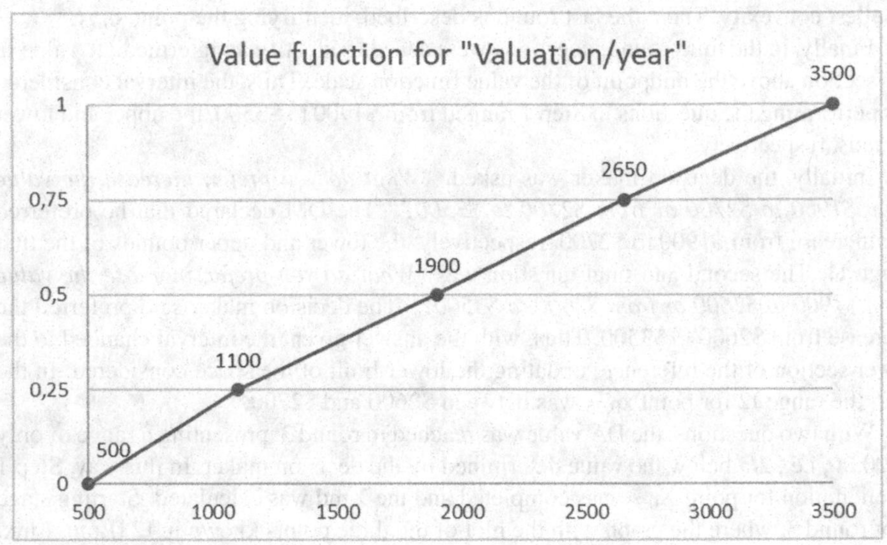

Fig. 7. Final graph with 'Valuation/year' criterion behavior.

5.1 Discussion

To compare performances, the same criterion was evaluated by the same decision maker, however using the traditional bisection method. That is, each question asked had the ultimate goal of making the decision maker declare a point of indifference between the compared values. Table 3 shows a comparison between the number of responses given with the "proposed approach" *versus* "bisection method", for each of the three intermediate points elicited.

Table 3. Comparison between traditional and adapted approaches

Approaches					
Proposed approach			Traditional bisection method		
Point	Number of questions	Final value	Point	Number of questions	Final value
$X_{0.25}$	3	$1100	$X_{0.25}$	7	$1095
$X_{0.5}$	4	$1900	$X_{0.5}$	8	$1825
$X_{0.75}$	2	$2650	$X_{0.75}$	5	$2675

Initial impressions reveal that, as expected, realize the intra-criterion elicitation using the traditional bisection method meant that the decision maker needed to answer a major number of questions compared to the approach proposed. Observing at point $X_{0.5}$, for example, it is possible to see that twice as many answers were necessary to obtain the final value. And when comparing the final values in both approaches, the bisection method with partial information differed by only $75 from that found in the traditional method.

For $X_{0.25}$ and $X_{0.75}$ references, the increase in the questions asked was greater than fifty percent. This clearly demanded more time to perform the procedure, as well as demands more cognitive effort on the part of the DM. Figure 8, presents the graph with the final behavior obtained with the traditional bisection method, where additional considerations can be explored.

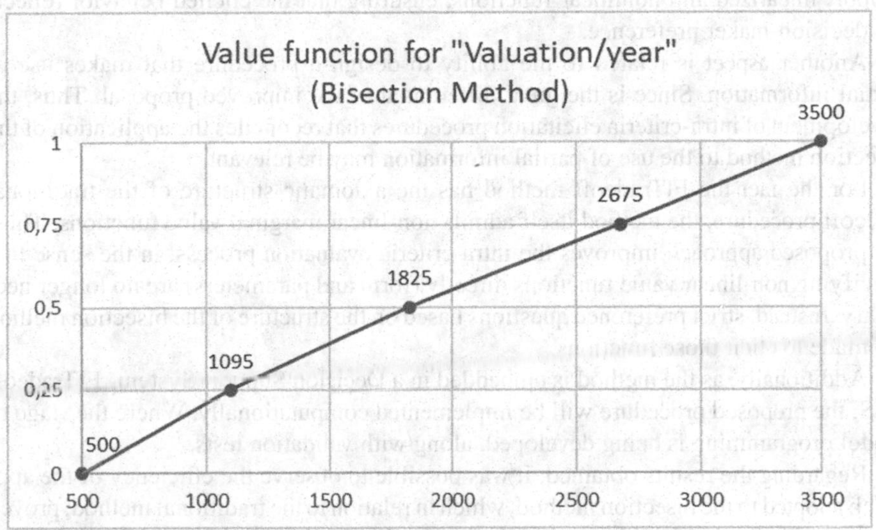

Fig. 8. Final graph using a traditional approach.

Analyzing Fig. 8, it is possible to verify that the final form identified through the elicitation process is visually similar to the behavior illustrated in Fig. 7. Including the final values found for each of the three points elicited. In $X_{0.25}$, for example, a difference of only $5 was identified.

Thus, establishing a parallel between the DA value declared by the decision maker in the elicitation using partial information, the difference between the $1100 found in the approach proposed and the $1095 obtained with the bisection method can be considered acceptable and consistent with the information provided by the decision maker.

Consequently, it was possible to verify that the bisection method with the use of partial information, proposed in the study, had a good performance, in a flexible process of elicitation. That presents advantages in terms of the effort and the time required and the structuring of the elicitation procedure.

6 Final Remarks

Initially, in opposition to models found in the literature, this study uses the performance values of the criteria of a multicriteria problem, to determine the space of admissible consequences. Defining a local measurement scale. Thus, the model is applied according to the circumstances, dynamically and seeking in fewer steps that the decision maker can express his preferences, using strict preference statements in a flexible procedure.

Once the intra-criteria evaluation in additive models consists in establishing the value function of each criterion, the proposal presented can be implemented in other methods that belong to this MCDM/A category. However, the axiomatic structure of these must support linearized and nonlinear functions, ensuring that the elicited behavior reflects the decision-maker preference.

Another aspect is related to the ability to design a procedure that makes use of partial information. Since is the great differential of the improved proposal. Thus, the development of intra-criteria elicitation procedures that reconciles the application of the bisection method to the use of partial information may be relevant.

For the fact the FITradeoff method has the axiomatic structure of the traditional tradeoff procedure, the method itself admits non-linear marginal value functions. Thus, the proposed approach improves the intra-criteria evaluation process, in the sense that specifying non-linear value functions directly (form and parameters) are no longer necessary. Instead, strict preference questions based on the structure of the bisection method are made to elicit those functions.

Additionally, as the method is embedded in a Decision Support System- FITradeoff DSS, the proposed procedure will be implemented computationally. Where the stage of model programming is being developed, along with validation tests.

Regarding the results obtained, it was possible to observe the efficiency of the approach adopted to the bisection method, which in relation to the traditional method, proved to be more agile, and less demanding in terms of cognitive effort.

For future research, other applications can be made considering improvements in the procedure, making the elicitation of discrete criteria also be included in the proposed approach. And for problems with a large number of evaluation criteria, it is interesting to investigate ways to reduce the number of criteria in the intra-criterion evaluation stage.

Acknowledgment. The authors are grateful for the support received and the funding provided by the Coordination for the Improvements of Higher Education Personnel (CAPES).

References

Belton, V., Stewart, T.: Multiple Criteria Decision Analysis: An Integrated Approach. Springer, Berlin (2002). https://doi.org/10.1007/978-1-4615-1495-4

Bertani, N., Boukhatem, A., Diecidue, E., Perny, P., Viappiani, P.: Fast and simple adaptive elicitations: Experimental test for probability weighting. Available at SSRN 3569625 (2020). https://doi.org/10.2139/ssrn.3569625

Chapman, J., Snowberg, E., Wang, S., Camerer, C.: Loss attitudes in the US population: Evidence from dynamically optimized sequential experimentation (DOSE). No. w25072. National Bureau of Economic Research (2018). https://doi.org/10.3386/w25072

da Cunha, C.P.C.B., de Miranda Mota, C.M., de Almeida, A.T., Frej, E.A., Roselli, L.R.P.: Applying the FITradeoff method for aiding prioritization of special operations of Brazilian federal police. In: de Almeida, A.T., Morais, D.C. (eds.) INSID 2020. LNBIP, vol. 405, pp. 110–125. Springer, Cham (2020). https://doi.org/10.1007/978-3-030-64399-7_8

De Almeida, A.T., Almeida, J.A., Costa, A.P.C.S., Almeida-Filho, A.T.: A new method for elicitation of criteria weights in additive models: flexible and interactive tradeoff. Eur. J. Oper. Res. **250**, 179–191 (2016). https://doi.org/10.1016/j.ejor.2015.08.058

de Almeida-Filho, A.T., de Almeida, A.T., Costa, A.P.C.S.: A flexible elicitation procedure for additive model scale constants. Ann. Oper. Res. **259**(1–2), 65–83 (2017). https://doi.org/10.1007/s10479-017-2519-y

De Almeida, A.T., Geiger, M., Morais, D.C.: Challenges in multicriteria decision methods. IMA J. Manag. Math. **29**(3), 247–252 (2018). https://doi.org/10.1093/imaman/dpy005

de Almeida, A.T., Roselli, L.R.P.: NeuroIS to improve the FITradeoff decision-making process and decision support system. In: Davis, F.D., Riedl, R., vom Brocke, J., Léger, P.-M., Randolph, A.B., Fischer, T. (eds.) NeuroIS 2020. LNISO, vol. 43, pp. 111–120. Springer, Cham (2020). https://doi.org/10.1007/978-3-030-60073-0_13

de Almeida, A.T., Frej, E.A., Roselli, L.R.P.: Combining holistic and decomposition paradigms in preference modeling with the flexibility of FITradeoff. CEJOR **29**(1), 7–47 (2021). https://doi.org/10.1007/s10100-020-00728-z

Edwards, W., Barron, F.H.: SMARTS and SMARTER: improved simple methods for multiattribute utility measurement. Organ. Behav. Hum. Decis. Process. **60**(3), 306–325 (1994). https://doi.org/10.1006/obhd.1994.1087

Eum, Y.S., Park, K.S., Kim, S.H.: Establishing dominance and potential optimality in multi-criteria analysis with imprecise weight and value. Comput. Oper. Res. **28**(5), 397–409 (2001). https://doi.org/10.1016/S0305 0548(99)00124-0

Fossile, D.K., Frej, E.A., da Costa, S.E.G., de Lima, E.P., de Almeida, A.T.: Selecting the most viable renewable energy source for Brazilian ports using the FITradeoff method. J. Clean. Prod. **260** (2020). https://doi.org/10.1016/j.jclepro.2020.121107

Frej, E.A., Roselli, L.R.P., de Almeida, J.A., de Almeida, A.T.: A multicriteria decision model for supplier selection in a food industry based on FITradeoff method. Math. Probl. Eng. **2017**, 1–9 (2017). https://doi.org/10.1155/2017/4541914

Frej, E.A., de Almeida, A.T., Costa, A.P.C.S.: Using data visualization for ranking alternatives with partial information and interactive tradeoff elicitation. Oper. Res. Int. J. **19**(4), 909–931 (2019). https://doi.org/10.1007/s12351-018-00444-2

Frej, E.A., Ekel, P., de Almeida, T.: Abenefit-to-cost ratio based approach for portfolio selection under multiple criteria with incomplete preference information. Inf. Sci. **545**, 487–498 (2021). https://doi.org/10.1016/j.ins.2020.08.119

Groothuis-Oudshoorn, C.G.M., Broekhuizen, H., van Til, J.: Dealing with uncertainty in the analysis and reporting of MCDA. In: Marsh, K., Goetghebeur, M., Thokala, P., Baltussen, R. (eds.) Multi-Criteria Decision Analysis to Support Healthcare Decisions, pp. 67–85. Springer, Cham (2017). https://doi.org/10.1007/978-3-319-47540-0_5

Jaszkiewicz, A., Słowiński, R.: The LBS-Discrete interactive procedure for multiple-criteria analysis of decision problems. In: Clímaco, J. (ed.) Multicriteria Analysis. Springer, Heidelberg (1997). https://doi.org/10.1007/978-3-642-60667-0_31

Kang, T.H.A., Frej, E.A., de Almeida, A.T.: Flexible and interactive tradeoff elicitation for multicriteria sorting problems. Asia Pacific J. Oper. Res. **37**(05) (2020). https://doi.org/10.1142/S0217595920500207

Keeney, R.L., Raiffa, H.: Decision Analysis with Multiple Conflicting Objectives. Cambridge University Press, New York (1976)

Lahdelma, R., Miettinen, K., Salminen, P.: Ordinal criteria in stochastic multicriteria acceptability analysis (SMAA). Eur. J. Oper. Res. **147**(1), 117–127 (2003). https://doi.org/10.1016/S0377-2217(02)00267-9

Narula, S.C., Vassilev, V.S., Genova, K.B., Vassileva, M.V.: A reference neighbourhood interactive method for solving a class of multiple criteria decision analysis problem. IFAC Proc. Volumes **37**(19), 131–137 (2004). https://doi.org/10.1016/S1474-6670(17)30671-7

Oliveira, G.D., Dias, L.C.: The potential learning effect of a MCDA approach on consumer preferences for alternative fuel vehicles. Ann. Oper. Res. **293**(2), 767–787 (2020). https://doi.org/10.1007/s10479-020-03584-x

Pergher, I., Frej, E.A., Roselli, L.R.P., de Almeida, A.T.: Integrating simulation and FITradeoff method for scheduling rules selection in job-shop production systems. Int. J. Prod. Econ. **227**, 107669 (2020). https://doi.org/10.1016/j.ijpe.2020.107669

Rodrigues, L.V.S., Casado, R.S.G.R., Carvalho, E.N.D., Silva, M.M.: Using FITradeoff in a ranking problem for supplier selection under TBL performance evaluation: an application in the textile sector. Production **30** (2020). https://doi.org/10.1590/0103-6513.20190032

Roselli, L.R.P., de Almeida, A.T., Frej, E.A.: Decision neuroscience for improving data visualization of decision support in the FITradeoff method. Oper. Res. Int. J. **19**(4), 933–953 (2019a). https://doi.org/10.1007/s12351-018-00445-1

Roselli, L.R.P., Pereira, L., da Silva, A., de Almeida, A.T., Morais, D.C., Costa, A.P.C.S.: Neuroscience experiment applied to investigate decision-maker behavior in the tradeoff elicitation procedure. Ann. Oper. Res. **289**(1), 67–84 (2019b). https://doi.org/10.1007/s10479-019-03394-w

Roselli, L.R.P., de Almeida, A.T.: The use of the success-based decision rule to support the holistic evaluation process in FITradeoff. Int. Trans. Oper. Res. (2021). https://doi.org/10.1111/itor.12958

Toubia, O., et al.: Dynamic experiments for estimating preferences: an adaptive method of eliciting time and risk parameters. Manage. Sci. **59**(3), 613 (2013). https://doi.org/10.1287/mnsc.1120.1570

Vasconcelos, G.R., de Mota, C.M.: Exploring multicriteria elicitation model based on pairwise comparisons: building an interactive preference adjustment algorithm. In: Mathematical Problems in Engineering, vol. 2019 (2019). https://doi.org/10.1155/2019/2125740

Zuheros, C., et al.: Sentiment analysis based multi-person multi-criteria decision making methodology using natural language processing and deep learning for smarter decision aid. Case study of restaurant choice using TripAdvisor reviews. Inf. Fusion **68**, 22 (2021). https://doi.org/10.1016/j.inffus.2020.10.019

Manufacturing Execution System Selection by Use of Multicriteria Partial Information Method

Jorge A. P. Mondadori[1(✉)], Mischel Carmen N. Belderrain[2], Rodrigo Jose Pires Ferreira[3], and Rafael V. Françozo[4]

[1] Serviço Nacional de Aprendizagem Industrial, Rua Belém, Centro, Londrina, PR, Brazil
[2] Instituto Tecnológico de Aeronáutica, Praça Marechal. Eduardo Gomes, 50, Vila das Acácias, São José dos Campos, SP, Brazil
[3] Universidade Federal de Pernambuco, Av. Prof. Moraes Rego, Recife, PE 1235, Brazil
rodrigo@cdsid.org.br
[4] Instituto Federal de Mato Grosso do Sul, Rua Pedro de Medeiros, Corumbá, MS 941, Brazil

Abstract. Seeking technologies to embrace Industry 4.0 has been a challenge to Small and Medium Enterprises (SME). Lack of investments and knowledge on tools and system integration put SMEs in a difficult position regarding their supply chain. As the main mission of the Industrial Federation of Paraná State (Fiep), serving industries to promote competitiveness and production optimization, one of the operation units, Artificial Intelligence Hub, developed a business model for a SaaS (System as a Service) solution for process digitalization in real time. Therefore, this paper focuses on choosing the most suitable online platform to integrate hardware and consulting services for online data acquisition and manufacturing execution. Decision making was made considering one internal alternative and four commercial alternatives being evaluated in a multicriteria model with nine criteria. In order to reduce cognitive efforts, a flexible and interactive procedure was implemented with possibility of partial information provided by Decision Maker. Findings from this work may provide a guideline for industrial operations and management leaders on choosing the right platform for process digitalization. It was also possible to guide the studied company (Fiep) on selecting software to help SMEs enter the Industry 4.0 scenario.

Keywords: Software selection · MAVT · FITradeoff · Partial information

1 Introduction/Purpose

Industry 4.0 implementation is a reality in every supply chain industrial sector. Among production activities, the main contributors to countries economic indicators are small and medium enterprises. Since they are generally part of a bigger production system, they must be connected to management systems (Moeuf et al. 2018). However, complex computer tools such MRP and ERP are expensive and rigid on their implementation, and small industries usually lack the specific knowledge for their implementation (Haseeb et al. 2019).

A. T. de Almeida and D. C. Morais (Eds.): INSID 2021, LNBIP 435, pp. 87–99, 2021.
https://doi.org/10.1007/978-3-030-91768-5_6

A simple way to aid SMEs to control their production is production indicators such as Overall Equipment Effectiveness. (OEE). OEE considers three factors (Wu et al. 2017):

1. Availability Efficiency – The ratio between Equipment up-time and Total time.
2. Performance Efficiency – The ratio between theoretical production time per unit and equipment up-time.
3. Quality Efficiency – The ratio between theoretical production time per effective unit and theoretical production time per unit.

Using automated data acquisition, these indicators can be built in real time.

Manufacturing Execution Systems (MES) are commonplace in bigger industries to control production and connect floor shop systems such as SCADA (Supervisory Control and Data Acquisition Systems) to ERP platforms (Haseeb et al. 2019). MES usually provides connectivity from planned production to manufacturing itself, controlling performance, availability and quality of equipment and produced goods, by use of sensors and hardware classified as IoT in the Industry 4.0 context (Chao and Li 2006; Batumalay and Santhapparaj 2009). These three indicators multiplied provide OEE.

The implementation of Industry 4.0 technologies in SMEs follows the choice of IoT systems and integration with analytics software, affecting business performance in a positive way, being considered as the most crucial introduction (Haseeb et al. 2019). It was also identified that SMEs ignore the growing number of management tools and technologies, since those industries lack investments in research and development activities (Moeuf et al. 2018).

The joint utilization of IoT and management software improves sustainability in economic, ecologic, and social aspects, considering competitiveness, time reduction, resource efficiency and human resources transformation. It also provides collection of large amounts of data that can be analyzed providing insights and tools for process optimization and implementation of lean manufacturing practices (Kiel et al. 2020).

Inside the industrial context, the Industrial Federation of Paraná State (Fiep) initiated activities on the first Artificial Intelligence Hub for Industries in Brazil, located in Londrina. The mission of the AI Hub is to provide access to artificial intelligence technologies for every industry. Considering small industries that lack relevant data and cannot afford services provided by expensive consultancy, AI Hub is developing a platform to act as an Industrial Virtual Assistant. In several workshops, SMEs pointed out the necessity of data acquisition and interpretation. These industries lack specific management knowledge regarding OEE and MES implementation.

The aim of this paper is to contribute to the field of data acquisition systems for small industries and OEE real time calculation by using the Flexible and Interactive Tradeoff (FITradeoff) method proposed by Almeida et al. (2016). The selection considered several criteria and alternatives, which were evaluated by the AI Hub to provide a cost efficient and easy way to implement the platform. The Multiattribute Value Theory (MAVT) was used to evaluate alternatives considered in the choice model, and the FITradeoff aided the decision maker to establish criteria scale constants (also called weight of criteria).

This paper is organized as follows: Sect. 2 discusses the application of multicriteria methods for choice and selection problems, as well as the general procedures of the FITradeoff method. The multicriteria model for platform selection is presented in Sect. 3.

Section 4 presents the application of the model developed in the previous section and its implication as results. Finally, conclusions and insights for future research are presented.

2 Background

2.1 Multicriteria Decision Models

In order to identify the best MES software for manufacturers, Chao and Li (Chao and Li 2006) implemented a model by use of the Analytical Network Process (ANP), developed by Saaty (Saaty 1996). Such method is conducted by successive pairwise comparisons among three alternatives and seven criteria. The authors found out that the best choice had already been considered by specialist decision maker, although pointing out that conducting the process of evaluation was a tiresome activity, demanding a lot of cognitive effort.

Comparing four multicriteria decision-making methods, Pätäri et al. (2018) discussed positive efficiency for portfolio selection in best-performing stocks. Different methods yielded similar investment returns corresponding to cumulative differences higher than 75% in comparison to traditional portfolio selection methods.

By using AHP (Analytical Hierarchy Process) (Saaty 1990), (Lai et al. 1999, 2002) identified best software technologies regarding multimedia authorizing systems, implementing the model on the AHP Expert Choice DSS (Decision Support System). Such systems aid decision makers to view the problem in a systemic way, visualizing different alternatives and criteria in a comparative form.

Multicriteria decision methods are divided in two big groups: non-compensatory (or outranking) and compensatory (or additive) (Almeida et al. 2015). The MAVT is usually used as the foundation for additive methods, such SMART, AHP and MACBETH (Montibeller et al. 2006). In these methods, a value function is used to aggregate a final value considering multiple criteria and alternatives.

In outranking methods, preference structures are used to compare criteria and alternatives. Unlike methods derived from MAVT, transitivity property may not be followed, since the decision maker's preferences may not agree with the logic demanded by additive methods. Outranking methods can be cited: ELECTRE and PROMETEE families (Govidan and Japsen 2015).

Multicriteria decision methods also have been used to aid reaching consensus in decision models (Lima et al. 2018), to solve facilities location problem under uncertainty (Niroomand et al. 2019), supplier segmentation together with fuzzy sets (Bai et al. 2017), for estimating warranty cost and production (Mitra and Patankar 1993), to mitigate supply risk with AHP and goal programming (Kull and Talluri 2008) and evaluate transportation projects performance in China (Shang 2004).

The FITradeoff (Almeida et al. 2016) method was developed in 2016, to overcome inconsistencies that occur with the traditional tradeoff procedure proposed by Keeney and Raiffa (1976). It is built based on the MAVT, which is an additive theory that is the foundation for other methods such as AHP, Macbeth, SMART. The main advantage pointed out in the method is that decision makers do not need to provide full information for the tradeoff procedure. The flexible elicitation minimizes the effort when making pairwise comparisons, then less inconsistency occurs (Palha 2016; Lima et al. 2017; Gusmao and Medeiros 2016; Frej et al. 2019).

2.2 FITradeoff Procedure

The FITradeoff method is supported by a Decision Support System (DSS), where decision maker (DM) may evaluate alternatives considering each criterion, without the need to evaluate criteria weights (Almeida et al. 2016). Then the value function is aggregated by using (1).

$$v(x) = \sum_{i=1}^{n} k_i v_i(x_i) \quad (1)$$

where:

i. x represents the vector of consequences of an alternative, considering all n criteria.
ii. k_i represents the scaling constant for the criterion i (also called weight of criterion).
iii. $v_i(x_i)$ represents the value function of the consequences x_i for the i criterion.

And, assuming (2).

$$\sum_{i=1}^{n} k_i = 1, \quad k_i \geq 0 \quad (2)$$

To apply the FITradeoff, it is first necessary to obtain the order of the weights by preference P relation. The first part yields the n-dimensional weight space:

$$\varphi_n = \left\{ (k_1, k_2, k_3, \ldots, k_n) \vee k_1 > k_2 > k_3 > \cdots > k_n; \ \sum_{i=1}^{n} k_i = 1; \ k_i \geq 0 \right\}$$

from the most relevant to the least relevant.

The DM does not need to define an exact value between consequences, just needs to choose between consequences. The DSS automatically calculates the upper and lower limit that the indifference can assume. For any criterion, relations (3) and (4) are established.

$$v_i\left(x_i'\right) > \frac{k_{i+1}}{k_i} \quad (3)$$

$$v_i\left(x_i''\right) < \frac{k_{i+1}}{k_i} \quad (4)$$

The DSS then classifies alternatives in three groups. The dominated alternatives are removed from the evaluation. The potentially optimal are chosen such as the elicitation method may continue. Optimal solutions represent the end of the method execution, presenting the best choice for the problem. This is represented in (5).

$$\max_{k_1, k_2, \ldots, k_n} \sum_{i=1}^{n} k_i v_i\left(x_{ij}\right), j = 1, 2, \ldots, m \quad (5)$$

It is then necessary to consider the LPP constraints to avoid strict inequality through the relations (3) and (4), and the constraint that yields the maximum value of the alternative j that should be greater than (or equal to) any other alternative in the subset (6).

$$\sum\nolimits_{i=1}^{n} k_i v_i(x_{ij}) \geq \sum\nolimits_{i=1}^{n} k_i v_i(x_{iz}),\ z = 1, 2, \ldots, m, z \neq j \tag{6}$$

The LPP then runs in order to find an optimal alternative. If not, the flexible elicitation process can start again. If desired. DM can see partial results or even not provide further information.

3 Proposed Model

As presented in the previous section, the objective of this paper is to select the best software for data acquisition, to be used as a platform in small industries, to provide adequate and real time information for production management. It is also desired to have the possibility of data gathering to develop an industrial virtual assistant, to aid the manager on their activity.

Since it is expected to be operated in the floor shop, the interface of hardware and software must be simple, with minimum setup steps, to avoid resistance usually encountered by automation systems.

Since such selection of software must be implemented by Fiep and its partners, it is important to establish criteria considering experience together with external experiences. The software has similar characteristics as MES software, and Chao (Chao and Li 2006) presented an architecture to solve this kind of problem.

The model proposed for this application is presented in Fig. 1. Steps 1 to 3 are considered data input to the system, while steps 4 and 5 are related to FITradeoff method.

In step 1, the facilitator and the decision maker must assess criteria discussed in literature review, followed by a justification of such criteria considering the real scenario of this application. It is also needed to identify in the market which commercial software alternative can be used to achieve the main goal, together with other possibilities such as internal development. Step 2 uses a MS Excel spreadsheet to evaluate each alternative considering every criterion. The DSS interprets the spreadsheet, normalizing data to be calculated in LPP. In Step 3, the decision maker establishes preferential order of criteria. Step 4 deals with the identification of criteria weight by executing the FITradeoff method. Step 5 presents results, that can be a set of potentially optimal alternatives or the maximized alternative as the best.

In the next section, the model presented in Fig. 1 is applied and each step of the method is discussed.

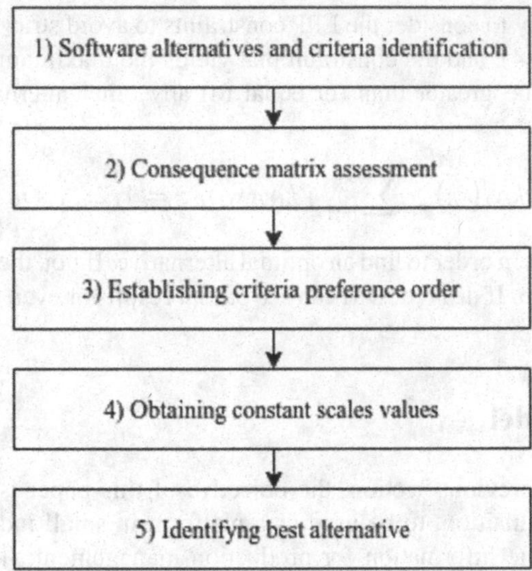

Fig. 1. Proposed model for software selection

4 Selection of a Software for Floor Shop Data Acquisition

For implementation, we identified five alternatives that could be used as efficient tools to provide real time data acquisition and information for SMEs. The first alternative (A1) is the possibility to allocate internal resources to develop the platform inside Fiep. Alternatives A2 to A5 are commercial solutions, defined by E1 (Enterprise 1) to E4 (Enterprise 4). Data for software evaluation were collected by industrial tests made by business consultants inside Fiep.

The authors of this paper acted as facilitator in the process, since the decision maker does not have knowledge in such methods. The decision maker is a technology and innovation manager inside the corporation.

Criteria were identified and represented functions that are continuous or discrete, as well may be maximized or minimized. Criteria characteristics are presented as follows:

C1 – Monthly Cost – Minimization continuous function – Represents monthly payment per machine per client.

C2 – Lifetime License – Maximization discontinuous binary function – Represents the possibility of having lifetime access to the platform.

C3 – Hardware Installation – Maximization continuous function – Represents ease of hardware installation by platform provider.

C4 – Support - Maximization continuous function – Represents quality of support that can be offered to clients.

C5 – Hardware Integration - Maximization continuous function – Represents possibility of integration with previously bought hardware.

C6 – System Maturity - Maximization continuous function – Represents the maturity of the platform and quantity of satisfied customers.

C7 – Database control – Maximization discontinuous function with five steps – Represents evaluation of data ownership by Fiep instead of supplier.

C8 – User Interface - Maximization continuous function – Represents ease of platform usage by floor shop workers.

C9 – Virtual Assistant Integration - Maximization continuous function – Represents easiness on connectivity to cognitive bot to be developed in further steps of AI Hub activities.

Consequences matrix of the problem is represented in Fig. 1, Step 2), and is presented in Table 1.

Table 1. Consequences matrix

	A1 Internal Dev.	A2 (E1)	A3 (E2)	A4 (E3)	A5 (E4)
C1	50	250	375	160	150
C2	1	0	0	1	0
C3	3	3	10	6	5
C4	4	10	6	4	3
C5	10	8	10	2	3
C6	2	8	5	8	6
C7	10	6	8	4	4
C8	2	10	10	9	7
C9	10	7	8	4	4

Following the model proposed, the spreadsheet was loaded in the FITradeoff DSS. The first step is to establish a criteria preference order to start the elicitation process. This preference establishes primary inequalities inside the LPP, and sets the sum of all weight equals to 1. This can be made holistically by pairwise comparison. The decision maker decided to choose his preference holistically, since all criteria are clear, independent, and well defined. Figure 2 is a screenshot of FITradeoff DSS, representing preference elicitation made by the decision maker.

Letter *P* represents "preference over next" and the chosen order for this case study is: C9 *P* C3 *P* C1 *P* C6 *P* C7 *P* C4 *P* C5 *P* C8 *P* C2. This means that the scaling constant of C9 is greater than the scaling constant of C3, which is greater than the scaling constant of C1 and so on. It also implies that the scaling constant of C9 is greater than every other criterion's scaling constant.

After establishing preference relations, FITradeoff DSS executes LPP for the first time. In this case, alternatives E1, E3 and E4 are considered dominated, and are eliminated from the process. Figure 3 shows partial numeric results. These partial results show the maximum value that each alternative can achieve inside the constant scale space. Specifically at this step A1 can achieve the maximum value, while A3 can achieve 0.8571. Both Internal Development and E2 are potentially optimal solutions for the LPP in at least one vector inside weight space.

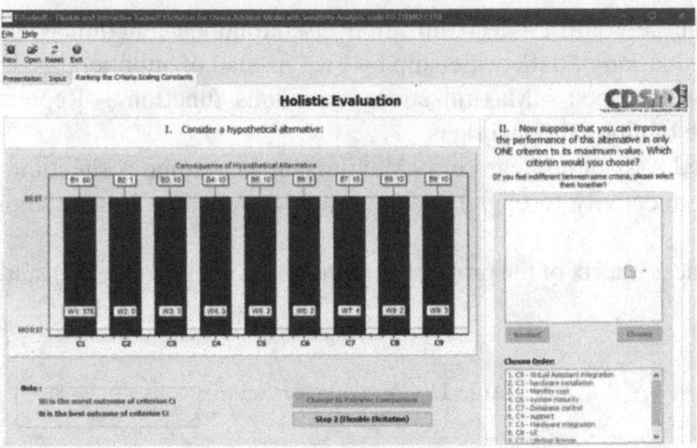

Fig. 2. Holistic preference relationship.

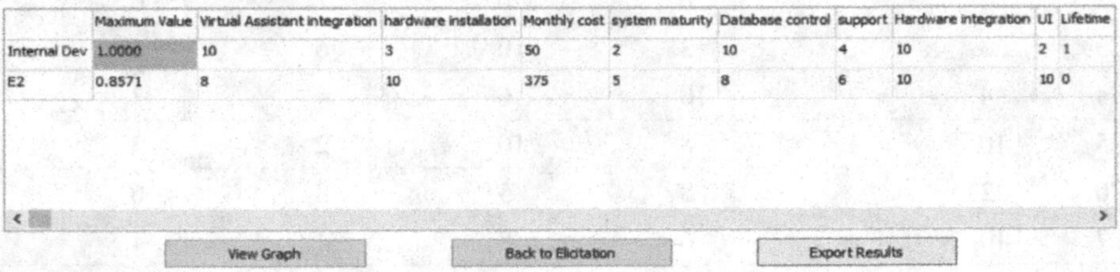

	Maximum Value	Virtual Assistant integration	hardware installation	Monthly cost	system maturity	Database control	support	Hardware integration	UI	Lifetime
Internal Dev	1.0000	10	3	50	2	10	4	10	2	1
E2	0.8571	8	10	375	5	8	6	10	10	0

View Graph | Back to Elicitation | Export Results

Fig. 3. Partial numeric results.

Still analyzing this first step, Fig. 4 represents each alternative performance over each criterion.

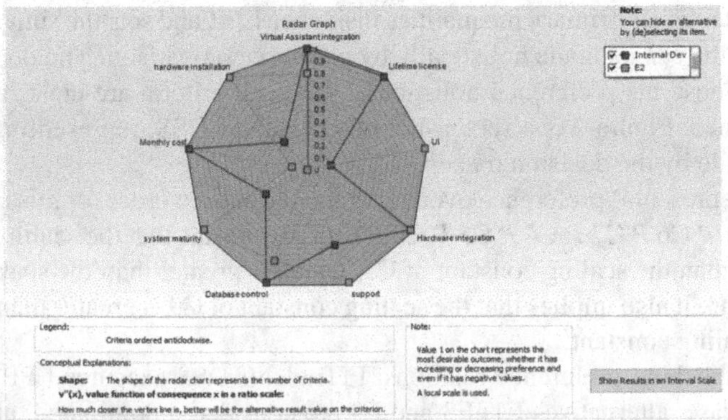

Fig. 4. Alternative performances over criteria.

Next step is the flexible elicitation process. In this step, the decision maker may choose between two consequences that compare two criteria. Each answer updates the space vector of LPP, by updating inequalities that act as restriction for the problem itself. In Fig. 5, the facilitator asked the following question to the decision maker, considering graphical interpretation: “What do you prefer: 0.5 performance on Virtual Assistant Integration (represented by blue bar) and the worst outcome on Lifetime license, or the best outcome on Lifetime License (green bar) and the worst outcome on Virtual Assistant Integration?”. Note that this is a hypothetical situation, that assigns the worst outcome to every other criterion, to make interpretation easier.

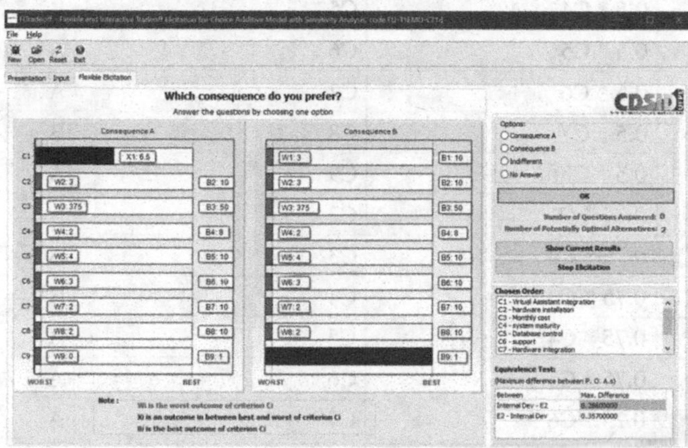

Fig. 5. First FITradeoff Cycle.

It is also important to keep track of answers given for each question made to the decision maker, to evaluate further results by replication of data. Table 2 presents every answer given by the decision maker until the end of the elicitation process.

After 17 answers, the decision maker decided to stop the elicitation process. As numeric partial results after flexible elicitation, A1 (Internal Development) could achieve a maximum value of 0.67 while A3 (E2) turned over numeric comparison achieving a maximum of 0.6747. Figure 6 represents minimum and maximum values that each scaling constant of each criterion could take. It was pointed out that the maximum difference is equal to 0.2 for criterion Virtual Assistant Integration.

It is also observed that there is no significant difference for criteria 7 to 9. This way, a sensitivity analysis was conducted by the facilitator and the decision maker, to test robustness of model implementation and evaluation.

A simulation module inside FITradeoff DSS provides sensitivity analysis. For this application, 10000 instances were executed, varying 10% around minimum and maximum values for criteria C1 to C6 and no variation for criteria C7 to C9, since their difference is insignificant.

Simulation for sensitivity analysis pointed out that for most cases the alternative E2 (A3) is the best solution, followed by Internal Development. Also, it is important to note that no entrant solution appeared, showing that these both alternatives are truly the best ones, as shown in Fig. 7.

Table 2. FITradeoff answer cycles

Cycle	Consequence A	Consequence B	Answer
1	0.5 * C1	C9	A
2	0.5 * C1	C2	B
3	0.5 * C2	C3	B
4	0.5 * C3	C4	B
5	0.5 * C4	C5	B
6	0.5 * C5	C6	A
7	0.5 * C6	C7	A
8	0.5 * C7	C8	B
9	0.5 * C8	C9	A
10	0.75 * C1	C2	A
11	0.75 * C2	C3	A
12	0.75 * C3	C4	B
13	0.75 * C4	C5	A
14	0.75 * C5	C6	A
15	0.25 * C6	C7	A
16	0.25 * C7	C8	B
17	0.75 * C8	C9	A

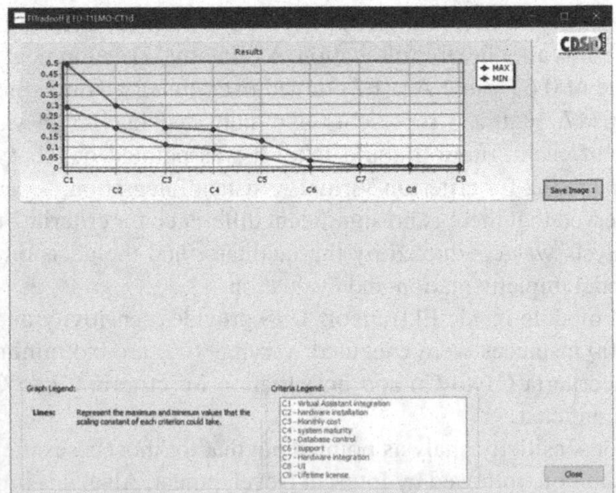

Fig. 6. Scaling constant final space vectors.

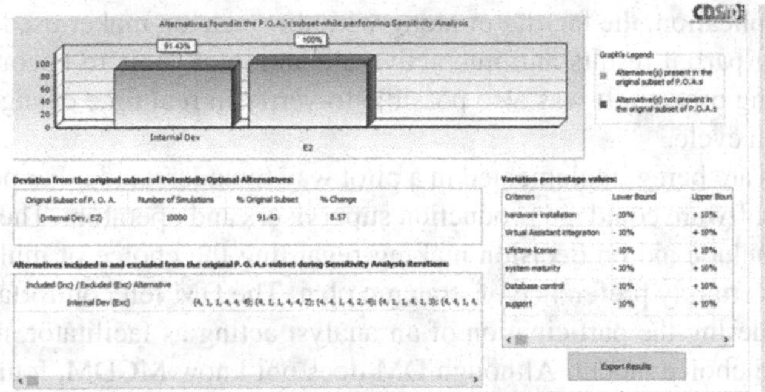

Fig. 7. Sensitivity analysis.

With better understanding of possible criteria scaling constant values, the decision maker asked for a second run, evaluating indifference among preference of criteria C1 to C6.

Letter I represents indifference, and the preference order for the second run is: C9 I C3 I C1 I C6 I C7 I C4 P C5 I C8 I C2. There is a preference relationship between C4 and C5 to separate the group identified as insignificant.

This time the LPP found A3 as the best solution without any flexible elicitation procedure.

Because of these results, the decision maker decided: he decided to allocate internal resources in the development of a solution. Since such development takes a lot of effort and usually needs to follow a long development cycle, alternative A3 is also considered.

Starting at the end of the implementation of the proposed model, both alternatives are considered and are being executed as pilots to test functional characteristics. This way, the software development may take small increments in proof of concept models to test data acquisition and functionality, while the alternative A3 provided fast industrial implementation, to validate both usage of MES and an alternative business model in caso of dead end of software development by any reason that may occur.

Software used for this application is version FU-T1EMO-CT1. There is a new and updated version of the FITradeoff DSS available at cdsid.org.br/fitradeoff.

5 Conclusion

Understanding its production efficiency is crucial for every SME that desires to increase competitiveness. Selection of software that supports such activities is important to provide decision insights in real time since FIEP's mission is to improve the production floor shop of its clients, decision making considering both quantitative and qualitative aspects are fundamental.

In this paper, we presented a model based on the FITradeoff method to aid our decision maker the possibility of flexible elicitation. Our model is validated by the selection of a MES system that will be used as a product together with the Industrial Virtual Assistant.

For our application, the facilitator along with the decision maker used the FITradeoff DSS, using partial results and interactive elicitation as tools to encourage a faster decision-making process. It was also possible to verify in real time changes on criteria weight for each cycle.

The results are being implemented in a pilot way to understand other possible needs that the MES software could aid production supervisors and operators. The implications of this work include aid on decision making regarding the choice of multiple alternatives under uncertainty preference of strategic plan. The DM felt comfortable using the method, considering the participation of an analyst acting as facilitator, mediating the knowledge and choice model. Although DM does not know MCDM, found the results consistent, the reason why DM chose to make a second evaluation described in Sect. 4, to try out the method performance. It was very important to guide the DM on the use of software, since he/she had no time to learn the foundations of applied methods.

It is also concluded that FITradeoff empowers decision makers that use additive models through easiness and data visualization. As future research, we suggest the update on Chao's (Chao and Li 2006) unified model using partial information, instead of ANP, to reduce decision maker cognitive effort in the scope of addressed problem, since it is an unified model and more industries are investing in MES.

Acknowledgment. The authors would like to acknowledge Capes and Fiep for partial financial support and the Center for Decision Systems and Information Development (CDSID) for providing the software.

References

de Almeida, A.T., de Almeida, J.A., Costa, A.P.C.S., de Almeida-Filho, A.T.: A new method for elicitation of criteria weights in additive models: flexible and interactive tradeoff. Eur. J. Oper. Res. **250**(1), 179–191 (2016)

de Almeida, A.T., Cavalcante, C.A.V., Alencar, M.H., Ferreira, R.J.P., de Almeida-Filho, A.T., Garcez, T.V.: Multicriteria and Multiobjective Models for Risk, Reliability and Maintenance Decision Analysis. ISORMS, vol. 231. Springer, Cham (2015). https://doi.org/10.1007/978-3-319-17969-8

Bai, C., Rezaei, J., Sarkis, J.: Multicriteria green supplier segmentation. IEEE Trans. Eng. Manage. **64**(4), 515–528 (2017)

Batumalay, K., Santhapparaj, A.S.: Overall equipment effectiveness (OEE) through total productive maintenance (TPM) practices—a study across the Malaysian industries. In: 2009 International Conference for Technical Postgraduates (TECHPOS), pp. 1–5 (2009)

Chao, L., Li, Q.: A unified decision model for evaluation and selection of MES software. In: Wang, K., Kovacs, G.L., Wozny, M., Fang, M. (eds.) PROLAMAT 2006. IIFIP, vol. 207, pp. 691–696. Springer, Boston (2006). https://doi.org/10.1007/0-387-34403-9_96

Frej, E.A., de Almeida, A.T., Costa, A.P.C.S.: Using data visualization for ranking alternatives with partial information and interactive tradeoff elicitation. Oper. Res. Int. J. **19**(4), 909–931 (2019). https://doi.org/10.1007/s12351-018-00444-2

Govidan, K., Jepsen, M.B.: ELECTRE: a comprehensive literature review on methodologies and applications. Eur. J. Oper. Res. **1**(29), 250 (2015)

de Gusmao, A.P.H., Medeiros, C.P.: A model for selecting a strategic information system using the FITradeoff. In: Mathematical Problems in Engineering (2016)

Haseeb, M., Hussain, H.I., Ślusarczyk, B., Jermsittiparsert, K.: Industry 4.0: a solution towards technology challenges of sustainable business performance. Soc. Sci. **8**(5), 154 (2019)

Keeney, R.L., Raiffa, H.: Decision Analysis with Multiple Conflicting Objectives. Wiley & Sons, New York (1976)

Kiel, D., Müller, J.M., Arnold, C., Voigt, K.I.: Sustainable industrial value creation: benefits and challenges of Industry 4.0. In: Proceedings of Digital Disruptive Innovation, pp. 231–270 (2020)

Kull, T.J., Talluri, S.: A supply risk reduction model using integrated multicriteria decision making. IEEE Trans. Eng. Manage. **55**(3), 409–419 (2008)

Lai, V.S., Trueblood, R.P., Wong, B.K.: Software selection: a case study of the application of the analytical hierarchical process to the selection of a multimedia authoring system. Inf. Manag. **36**(4), 221–232 (1999)

Lai, V.S., Wong, B.K., Cheung, W.: Group decision making in a multiple criteria environment: a case using the AHP in software selection. Eur. J. Oper. Res. **137**(1), 134–144 (2002)

Lima, A.S., de Souza, J.N., Moura, J.A.B., da Silva, I.P.: A consensus-based multicriteria group decision model for information technology management committees. IEEE Trans. Eng. Manage. **65**(2), 276–292 (2018)

Lima, E.S., Viegas, R.A., Costa, A.P.C.S.: A multicriteria method-based approach to the BPMM selection problem. In: 2017 IEEE International Conference on Systems, Man, and Cybernetics (SMC), pp. 3334–3339. IEEE (2017)

Mitra, A., Patankar, J.G.: An integrated multicriteria model for warranty cost estimation and production. IEEE Trans. Eng. Manage. **40**(3), 300–311 (1993)

Moeuf, A., Pellerin, R., Lamouri, S., Tamayo-Giraldo, S., Barbaray, R.: The industrial management of SMEs in the era of Industry 4.0. Int. J. Prod. Res. **56**(3), 1118–1136 (2018)

Montibeller, G., Gummer, H., Tumidei, D.: Combining scenario planning and multi-criteria decision analysis in practice. J. Multi Criteria Decis. Anal. **14**, 5–20 (2006)

Niroomand, S., Mosallaeipour, S., Mahmoodirad, A.: A hybrid simple additive weighting approach for constrained multicriteria facilities location problem of glass production industries under uncertainty. IEEE Trans. Eng. Manage. **67**(3), 846–854 (2019)

Palha, R.P., de Almeida, A.T., Morais, D.C.: Plataforma De Negociação Com Elicitação De Preferências Através Do Fitradeoff. In: Proceedings of XLVIII SBPO - Simpósio Bras. Pesqui. Operacional, pp. 4088–4092. SOBRAPO (2016)

Pätäri, E., Karell, V., Luukka, P., Yeomans, J.S.: Comparison of the multicriteria decision-making methods for equity portfolio selection: the US evidence. Eur. J. Oper. Res. **265**(2), 655–672 (2018)

Saaty, T.L.: How to make a decision: the analytic hierarchy process. Eur. J. Oper. Res. **48**(1), 9–26 (1990)

Saaty, T.L.: Decision Making with Dependence and Feedback: The Analytic Network Process: The Organization and Prioritization of Complexity. RWS Publications, Pittsburgh (1996)

Shang, J.S., Tjader, Y., Ding, Y.: A unified framework for multicriteria evaluation of transportation projects. IEEE Trans. Eng. Manage. **51**(3), 300–313 (2004)

Wu, Y.H., Wang, S.D., Chen, L.J., Yu, C.J.: Streaming analytics processing in manufacturing performance monitoring and prediction. In: 2017 IEEE International Conference on Big Data (Big Data), pp. 3285–3289. IEEE (2017)

Incorporating Hierarchical Criteria Structure in the Fitradeoff Method

Maria Júlia Leal Vieira[1(✉)], Eduarda Asfora Frej[1,2], Adiel Teixeira de Almeida[1,2], and Francisco Filipe Cunha Lima Viana[1]

[1] Department of Production Engineering, Graduate Program in Management Engineering, Federal University of Pernambuco (UFPE), Architecture Avenue, s/n, University City, Recife, PE, Brazil
mariajulia.vieira@ufpe.br, {eafrej,almeida}@cdsid.org.br
[2] Center for Decision Systems and Information Development – CDSID, University Federal of Pernambuco (UFPE), Architecture Avenue, s/n, University City, Recife, PE, Brazil

Abstract. Multicriteria Decision Making/Aiding (MCDM/A) techniques are usually required to solve practical decision-making problems that consider multiple criteria structured based on a value tree. Structuring criteria based on a hierarchy is common specially in problems in which the number of criteria is high, and therefore MCDM/A techniques should be prompted to deal with such situations. The well-known FITradeoff method is being widely applied for solving practical multicriteria problems due to its easiness of use and attractive flexibility features. However, the current version of this method is suitable for dealing with single-level criteria decision problems only. Therefore, in this context, this paper proposes a approach for solving multicriteria decision-making problems with hierarchically structured criteria in the FITradeoff method. This approach uses partial information of preferences provided by the decision maker, based on a structured process within the scope of the multi-attribute value theory, to find the values of the scale constants. The model is presented for both choice and ranking problematics and it is based on the traditional tradeoff procedure, which is axiomatically robust. The model effectiveness is verified after being applied to three problems adapted from the literature to both choice and ranking problematics. As a result, it was observed that in the choice problematic, in all analyzed problems, a single optimal alternative was found and always with 6 or less questions answered. In turn, in the ranking problematic in all cases either a complete order or a complete preorder was found with 17 or less questions answered.

Keywords: Hierarchical criteria · Multiple Criteria Decision-Making Aiding (MCDMA) · FITradeoff

1 Introduction

Decision situations typically present conflicting objectives, dynamic environments or poorly structured problems (Louvieris et al. 2010). In this context, the identification and structuring of objectives presents an enormous potential to support decision making,

A. T. de Almeida and D. C. Morais (Eds.): INSID 2021, LNBIP 435, pp. 100–118, 2021.
https://doi.org/10.1007/978-3-030-91768-5_7

since they facilitate the understanding of the problem under analysis (Keeney 1992). However, there are still some difficult to understand the elements of decision structuring such as the differences between ends and means objectives as well as objectives and goals, restrictions, or even alternatives. Thus, the relations between the elements are often not properly specified (Keeney 1996).

The identification of these objectives, in each analyzed problem, requires time, creativity, and knowledge of the decision-makers, who must be assisted by analysts during the process (Keeney and Raiffa 1993). Given the above, decision-aiding tools require concepts and methods from mathematical origin or simpler ways of organizing thought, such as lists, tree structures, or graphs (Boyssou et al. 2002).

Thus, the front end of formal assessment models and the first step in any of its applications is the elicitation or construction of a formal value structure, usually in the form of trees of values and objectives. Among these formal value structures, a hierarchy of objectives stands out (Keeney and Raiffa 1976), which organizes objectives and attributes to clarify decision making, also known as the value tree or decision hierarchy (Von Winterfeldt and Edwards 1986).

The hierarchical structure has several advantages. First, understanding the values established for the weights of the objectives, which results in a better elicitation process. In addition, the hierarchy makes clearer the distinction between means and ends objectives, identifying gaps, redundancy, or double counts (Keeney 1992). However, the formulation of the value tree is not an easy task and takes up most of the time in real applications. Since these structures should include all relevant aspects, but still be as small as possible. Also, an important aspect is that for additive value models attributes of the value tree should be preferably independent of each other (Poyhonen et al. 2001).

In this case, defining criteria weights values is not a trivial task since these parameters should not only represent the level of importance of the criteria but represent the meaning of substitution rates, called scale constants. Thus, decision-makers should evaluate criteria considering how much they are willing to lose in one criterion to win in another one (Frej et al. 2021).

It is worth noting that in the last decades, several authors are faced the challenge of inferring partial information in hierarchical structures. Due to the difficulty in determining these weights, it was found that biases are generated at the time of elicitation, but that their origins are not exactly known (Poyhonen 1998). Aiming to determine the weights in value tree structures and to minimize or eliminate these biases, multicriteria decision support methods that deal with hierarchies were developed, such as the AHP (Analytical Hierarchy Process) (Saaty 1980), PAIRS (Preference Assessment by Imprecise Ratio Statements) (Salo and Hamalainen 1992), PRIME (Preference Ratios in Multiattribute Evaluation) (Salo and Hamalainen 2001) and RICH (Rank Inclusion in Criteria Hierarchies) (Salo and Pukka 2005). However, these still have some limitations and inconsistencies, such as the AHP which has several disadvantages, as presented by Belton and Goodwin (1996).

In this context, it is a challenge to analyze hierarchically structured criteria in the context of partial information highlighting the need to research new methods, since many of them are effective in solving problems, but do not accept hierarchically organized criteria as input. An example is the FITradeoff method (Flexible and Interactive Tradeoff), which deals directly with the objectives present at the bottom of the hierarchy, not being capable to infer information about the lower levels criteria through the objectives at the highest level. Therefore, this work aims to propose a new approach to consider hierarchical criteria within the FITradeoff method, increasing the applicability of the method to solve problems with hierarchically structured objectives and, possibly, in these cases, reducing the number of questions answered to obtain the solution of the problem.

This work is structured as follows. Section 2 describes the concept of decision tree, Sect. 3 describes the FITradeoff method both for choice and rank problematics, Sect. 4 presents the new approach to hierarchical problems, Sect. 5 presents the results and discursion. Finally, Sect. 6 presents the final comments and lines for futures research.

2 Criteria Hierarchy

A fundamental step towards the solution of a problem is the definition of which criteria will be considered for the evaluation of the alternatives. An objective, generally, indicates the direction to be followed to obtain better results, and it is measured according to these attributes. In this way, if the decision-maker neglects one of the main objectives, then the information that could be used to distinguish the alternatives can be ignored, increasing the likelihood of an error at the time of the decision (Keeney and Raiffa 1993).

Given this, the structuring of objectives results in a process of deeper and more precise understanding of what should be considered in the context of the decision, making it clear and defining the set of final objectives (Keeney 1996). This is usually done by obtaining the criteria through a hierarchical construction in the form of a value tree. In this regard, the ends objectives, relatively broader, are represented at the higher level and are increasingly divided into more specific criteria (Belton and Stewart 2002). On the other hand, the lower-level criteria must be mutually exclusive and collectively provide an exhaustive characterization of the higher-level criteria to include all the fundamental aspects of the consequences of the decision alternatives and to avoid double counting (Kajanus et al. 2004).

Analysis on value trees is usually based on additive value models, and, consequently, weights depend on the range of attributes and should be normalized. In these cases, elicitation can be carried out hierarchically or non-hierarchically, as shown in Fig. 1 (Poyhonen et al. 2001). In non-hierarchical weighting, the decision-maker considers all the attributes of the lower level simultaneously and assigns weights only to them. In turn, in the hierarchical weighting, the weights of each level and each branch of the value tree are elicited and normalized to sum equals one, separately. In this way, the final weights of the lowest hierarchical level are obtained by multiplying all the weights in the value tree (Weber and Borcherding 1993).

Regardless of how the analysis of the structure is made, the final results should not be modified, if there is a change it may be indicative of the existence of biases at the time of determining the weights. Another important aspect was stated by Belton and Stwart (2002) that warned for the necessity of defining the difference between cumulative and relative weights, in italics and in bold in Fig. 2, respectively. The relative weights are evaluated inside families of criteria being normalized to sum equals one, that is, criteria that share the same objective at the above level, as, in Fig. 2, US accessibility and quality

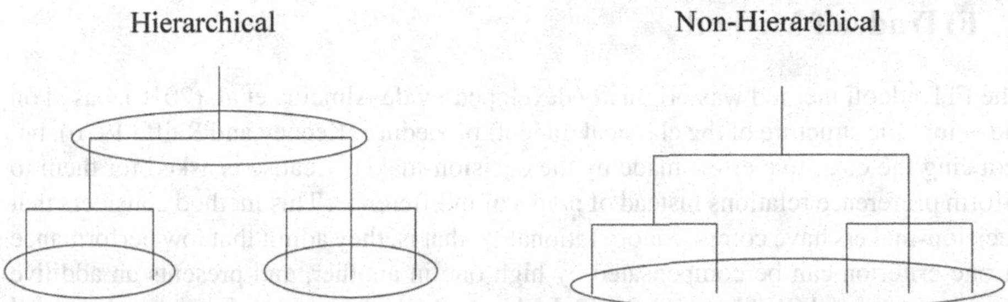

Fig. 1. Representation of the hierarchical and non-hierarchical form of determining weights. (Source: Adapted from Poyhonen et al. (2001))

of life. In turn, the cumulative weights of a criterion are the product of its relative weight and the relative weight of the objective at the above level that is associated with this, and so on up to the top of the tree, for example, in Fig. 2, the multiplication of the relative weight of the criterion personal problem by the relative weight of the criterion staff availability results in the cumulative weight of the criterion staff availability.

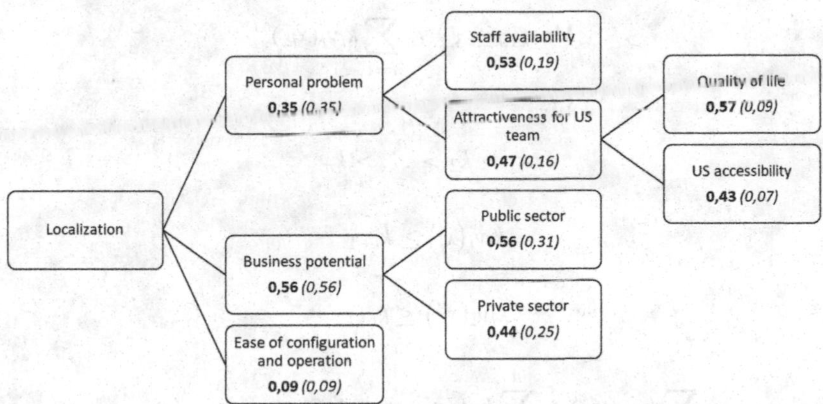

Fig. 2. Representation of relative and cumulative weights. (Source: Belton and Stewart (2002))

The authors state that higher-level criteria weights are harder to interpret since they are the sum of their sub-criteria cumulate weights. Other authors highlighted the difficulty of defining values for the scale constants in value trees and warned of several biases that may originate from this process (Poyhonen et al. 2001). Thus, the definition of scaling constants is a field with potential to be explored, especially in multicriteria decision problems, since the opinion of the decision-maker needs to be taken into consideration for its determination. In this way, the present work presents a mathematical model that allows the definition of these values through a hierarchical elicitation, that will be incorporated in the FITradeoff system.

3 FITradeoff Method

The FITradeoff method was originally developed by de Almeida et al. (2016), based on the axiomatic structure of the classical tradeoff procedure (Keeney and Raiffa 1976), but reducing the cognitive effort made by the decision-maker because is asked for them to inform preference relations instead of points of indifference. This method considers that decision-makers have compensatory rationality, that is, they admit that low performance in one criterion can be compensated by high one in another, and presents an additive aggregation model (Pergher et al. 2020). In this method, after structuring the problem and defining alternatives and criteria, an intra-criterion evaluation should be carried out, that is, converting the consequences of the defined criteria into a single 0–1 scale. This step ensures that a global value for each alternative is specified through additive aggregation at the end of the process (Belton and Stwart 2002). Then, an inter-criterion analysis is carried out, in which the flexible and iterative Tradeoff presents two main stages: ranking the weights of the criteria and eliciting their values (Roselli et al. 2018).

In the case of the choice problematic, after the decision-maker has established his preference, a linear programming problem (LPP) (Eq. 1 to 7) is executed to verify the potential optimality of the alternatives (De Almeida et al. 2016).

$$Max_{k1,k2,k3,\ldots,kn} \sum_{i=1}^{n} k_i v_i(x_{ia}) \tag{1}$$

$$s.t.$$

$$k_1 > k_2 > \ldots > k_n \tag{2}$$

$$k_i v_i\left(x_i'\right) \geq k_{i+1} \tag{3}$$

$$k_i v_i\left(x_i''\right) \leq k_{i+1} \tag{4}$$

$$\sum_{i=1}^{n} k_i v_i(x_{ia}) \geq \sum_{i=1}^{n} k_i v_i(x_{iz}),\ z = 1, 2, \ldots, m;\ a \neq z \tag{5}$$

$$\sum_{i=1}^{n} k_i = 1 \tag{6}$$

$$k_i \geq 0,\ i = 1, 2, 3, \ldots, n \tag{7}$$

In this LPP, k_i is the scale constant from the i criterion and the decision variables. Furthermore, the model is applied for each alternative *a*, in the set of m alternatives, considering consequences x_{ia} for criterion i, in the set of n criteria, and alternative *a*. Being $v_i(x_{ia})$ the value function of the consequence x_{ia} normalized in a 0 to 1 scale.

Equation (1) is the objective function of the LPP which seeks to maximize the value of alternative *a*. Then, the constraints of the linear programming are defined as follows: (2) ranking of the scale constants, obtained in the first stage; (3) and (4) restrictions resulting from the strict preferences established by the decision-maker; (5) potential optimality constraint, which aims to ensure that the global value of the alternative a is greater than the global value of the alternative z, $a \neq z$ for at least one weight vector; and (6) and (7) normalization and non-negativity of scale constants, respectively (De Almeida et al. 2016).

In the case of the ranking problematic, the objective function of the LPP and the restrictions are modified, since it is desired to rank the alternatives. Thus, the restriction of potential optimality does not belong to the LPP, once it is not desired to exclude alternatives from the final result (Frej et al. 2019). The LPP for the ranking problematic is represented by Eq. 8 to Eq. 13.

$$MaxD(a_a, a_z) = \sum_{i=1}^{n} k_i v_i(x_{ia}) - \sum_{i=1}^{n} k_i v_i(x_{iz}) \tag{8}$$

$s.t.$

$$k_1 > k_2 > \ldots > k_n \tag{9}$$

$$k_i v_i(x_i') \geq k_{i+1} \tag{10}$$

$$k_i v_i(x_i'') \leq k_{i+1} \tag{11}$$

$$\sum_{i=1}^{n} k_i = 1 \tag{12}$$

$$k_i \geq 0, i = 1, 2, 3, \ldots, n \tag{13}$$

The LPP referring to the choice problematic aims to use the concept of potential optimality to find an optimal alternative to the problem under analysis, while the LPP for the ranking problematic uses the concept of pairwise dominance to find the rank of alternatives (Frej et al. 2019).

In the FITradeoff decision support system, the process begins with the stage of creating a ranking of the criteria weights, being this the initial information of preference provided by the decision-maker (Pergher et al. 2020). Next, there is a comparison of hypothetical consequences that should be conducted by the decision-maker, establishing strict preference or indifference for each compared pair. It is also possible to choose to not respond to the comparison of a certain cycle (De Almeida et al 2016). During this phase, called elicitation, the partial results can be monitored through radar, bubble, and bar graphs that allow a better understanding of the alternatives under analysis. In addition, at each cycle, the decision-maker can choose to continue the process or stop and remain with the result found up to that moment (Roselli et al. 2018).

If the decision-maker decides to continue the process until it reaches the end, for the choice problematic one or more alternatives can be obtained as optimal for the problem, for the ranking problematic a complete or partial rank can be obtained (Frej et al. 2019). This method is previously applied to solve problems in the most diverse areas, such as location problems (Dell'ovo et al. 2020), supplier selection (Rodrigues et al. 2020; Frej et al. 2017), portfolio selection (Frej et al. 2021), scheduling rules selection in job-shop production systems (Pergher et al. 2020), renewable energy source selection for Brazilian ports (Fossile et al. 2020) and selection of an agricultural technology package (Carrillo et al. 2018).

In addition to the various applications, FITradeoff has recently started to allow the realization of holistic evaluations linked to evaluations by decomposition, which are the two basic paradigms of preference modeling in MAVT (de Almeida et al. 2021). However, in none of these applications the hierarchical structured criteria were considered, which shows a limitation of the current method that only analyzes criteria present at the base of the value tree. Therefore, this work aims to propose a new mathematical model to allow the resolution of problems in which the criteria are hierarchically structured in FITradeoff.

4 Incorporating Hierarchically Structured Criteria to the FITradeoff Method

The new approach to hierarchical problems uses the basic concepts of the actual FITradeoff for choice problematic, potential optimality, and for ranking problematic, dominance relations. However, the consequences are associated with the criteria located at the base of the hierarchy originated from a certain criterion located at the top of it, instead of associated only with the criteria located at the base of the hierarchy. Thus, in the choice problematic, the LPP is applied to each alternative *a*, considering the consequences x_{ija} where *i* represents the higher-level criterion and *j* the lower-level criterion of the hierarchy.

On the other hand, in the ranking problematic, the concept of pairwise dominance is considered, in which an alternative *a* dominates another alternative *z* if and only if the global value of *z* cannot be greater than the global value of *a* for any vector of weights within the weight space Φ, considering, therefore, consequences x_{ija} and x_{ijz} where *i* represents the higher-level criterion and *j* the lower-level criterion of the hierarchy.

Thus, before presenting the mathematical model it is important to highlight some assumptions and properties to support the development of the proposed approach. Figure 3 presents a hierarchical representation of the criteria that are considered in the model proposed in this paper. In Fig. 3, 1, 2 … n are the objectives located in the higher level that are subdivided into objectives located in the lower level, and m_n is the number of criteria belonging to the family of the higher-level criterion n. Additionally Fig. 3 serves as a basis for the assumptions and properties presented.

Thus, k_{ij} is the weight of the criterion, in which *i* designates the higher-level criterion, in this case, i = 1, 2, …, n − 1, n and *j* designates the lower-level criterion. For example, k_{12} is the weight of criterion 2, located in the lower level, who is associated with the criterion 1 located in the higher level.

In addition to the information about value trees that were introduced in Sect. 2, it is important to present some properties of the cumulative and relative weights. As defined above, the relative weights are determined by evaluations within families of criteria,

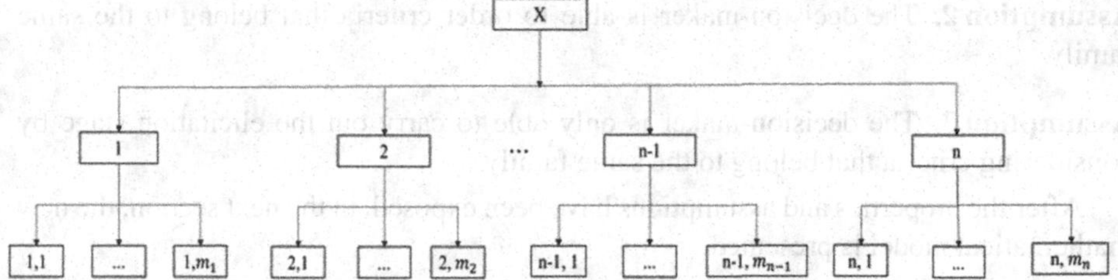

Fig. 3. Hierarchical representation.

which need to sum equals one. In turn, the cumulative weight of a criterion is the product of its relative weight and the relative weight of higher-level criterion associated, and so on up to the top of the value tree. Assuming a two-level hierarchy, Eq. 14 represents the cumulative weight calculation.

$$k'_{ij} = k_i.k_{ij} \tag{14}$$

These cumulative weights have three characteristics that are important to highlight. The first characteristic is that the cumulative weights of the criteria in the higher level are the sum of the cumulative weights of their associated criteria in the lower level. Equation 15 represents this sum.

$$\sum_{j=1}^{m_i} k'_{ij} = k'_i; \forall i, i = 1, \ldots, n \tag{15}$$

The second characteristic is that if the criterion is located at the top of the value tree, then it presents the cumulative weight equal to the relative weight, as represented in Eq. 16.

$$k'_i = k_i \tag{16}$$

Lastly, the third characteristic is that if the criterion in the higher level is not subdivided then the cumulative weight from that criterion is considered in the sum of the cumulative weights of the criteria in the lower level belonging to the other families, so that the sum of all cumulative weights in one level is 1.

Assumption 1. A two-level hierarchy is considered and the weights of the criteria located at the top of the hierarchy are determined by: ROC weights (Rank Ordered Centroid) (Edwards and Barron 1994) or it can be defined directly.

This assumption was made to allow information about the values of the scale constants of the lower-level criteria to be inferred from the higher-level ones. Thus, if a range of values was established for the scale constants of the higher-level criteria, more attention would be needed by the decision-maker to prevent intercepts between the intervals, increasing the cognitive effort to define the range limits. Furthermore, in future works, when hierarchies of more than two levels are analyzed, the mathematical programing model would become non-linear, which increases the complexity of the problem, since several intervals would need to be considered to infer the information. Therefore, by means of simplification, this assumption was established.

Assumption 2. The decision-maker is able to order criteria that belong to the same family.

Assumption 3. The decision-maker is only able to carry out the elicitation stage by considering criteria that belong to the same family.

After the properties and assumptions have been exposed, in the next section, the new mathematical model is presented.

4.1 Mathematical Model

The actual FITradeoff presents an elicitation stage in which all the criteria can be compared with each other using a non-hierarchical elicitation. However, it is important that the method enables the resolution of problems that present criteria structured in value trees, through a hierarchical elicitation. In this case, decision-makers are able to answer questions by comparing the criteria, located at the base of the value tree, which belong to the same family, but not between criteria from different families.

This issue emphasizes the need to establish relations of pairwise dominance in the ranking problematic, and of potential optimality in the choice problematic, using these elicitation characteristics that are different from the applied in the actual FITradeoff. Therefore, a new LPP model should be executed after each answer given by the decision-maker to find a set of potentially optimal alternatives or the ranking of alternatives, depending on the chosen problematic. Thus, this new mathematical model was developed to solve problems where the criteria are structured in a two-level hierarchy and the following LPP, represented by Eq. 17 to 27, is performed for a choice problematic.

$$\max_{k_{11},\ldots,k_{21},\ldots,k_{n,m_n}} \sum_{i=1}^{n}\sum_{j=1}^{m_i}(k_i.k_{ij}).v_{ij}\left(x_{ija}\right) \tag{17}$$

$s.t.$

$$\sum_{i=1}^{n} k_i = 1 \tag{18}$$

$$\sum_{j=1}^{m_i} k_{ij} = 1, \forall i, i = 1, \ldots, n \tag{19}$$

$$k_1 > k_2 > \ldots > k_{n-1} > k_n \tag{20}$$

$$k_{i1} > k_{i2} > \ldots > k_{i,j+1} > \ldots > k_{i,m_i} \tag{21}$$

$$\begin{aligned} &k_{ij}.v_{ij}\left(x'_{ij}\right) \geq k_{ij+1} - \varepsilon; j = 1, 2, \ldots, m_i \\ &k_{ij}.v_{ij}\left(x''_{ij}\right) \leq k_{ij+1} - \varepsilon; j = 1, 2, \ldots, m_i \end{aligned} \tag{22}$$

$$\sum_{j=1}^{m_i} k_i.k_{ij} = k'_i; \forall i, i = 1, \ldots, n \tag{23}$$

$$\sum_{j=1}^{m_n}\sum_{i=1}^{n} k'_{ij}.v_{ij}(x_{ija}) \geq \sum_{j=1}^{m_n}\sum_{i=1}^{n} k'_{ij}.v_{ij}(x_{ijz}); z = 1, 2, \ldots, w; z \neq a \quad (24)$$

$$k'_i > k'_{i+1} \therefore \sum_{j=1}^{m_i} k_i.k_{ij} > \sum_{j=1}^{m_{i+1}} k_{i+1}.k_{(i+1)j} \quad (25)$$

$$k'_i \geq 0; i = 1 \; to \; n \quad (26)$$

$$k'_{ij} \geq 0; i = 1 \; to \; n; j = 1 \; to \; m_i \quad (27)$$

In this LPP, k_i is the relative scale constant from the higher-level i criterion, in the set of n higher-level criteria, k_{ij} is the relative scale constant from the lower-level j criterion, in the set of m_i lower-level criteria, associated with the higher-level i criterion, and k_{ij} is also the decision variables. Consequently, k'_i is the cumulative scale constant from the higher-level i criterion, in the set of n higher-level criteria, and k'_{ij} is the cumulative scale constant from the lower-level j criterion, in the set of m_i lower-level criteria, associated with the higher-level i criterion. Furthermore, the model is applied for each alternative a, in the set of w alternatives, considering consequences x_{ija} for the lower-level criterion j associated with the higher-level criterion i, and alternative a. Being $v_{ij}(x_{ija})$ the value function of the consequence x_{ija} normalized in a 0 to 1 scale.

Analyzing the LPP, the first Eq. (17) is the objective function and seeks to maximize the global value of the alternative a. The relative weights of the lower-level criterion (k_{ij}) are the decision variables. In turn, the relative weight of the higher-level criterion (k_i) will in principle be a fixed value that can be defined from three different forms, as described in assumption 1. Furthermore, it is important to highlight that if the higher-level criterion is not subdivided then in the objective function would be enough to multiply the value function of alternative a $(v_{ij}(x_{ija}))$ by the relative weight of this criterion (k_i).

Then, the restrictions of the LPP are observed: (18) and (19) they are the normalization of the relative weights of the higher-level criteria and the normalization of the relative weights of the lower-level criteria that belong to the same family, respectively; (20) ranking of the higher-level criteria; (21) ranking of the lower-level criteria that belonging to the same family; (22) refers to the elicitation of preferences of the relative weights of the lower-level criteria that belonging to the same family; (23) guarantees the condition of Eq. 15, however, care must be taken as it does not need to be used in the case where only two hierarchical levels exist, since it becomes similar to constraint 19. This fact occurs because, in this case the Eq. 16 is applied, allowing a simplification of this equation; (24) potential optimality constraint, which aims to ensure that the global value of the alternative a is greater than the global value of the alternative z, $a \neq z$ for at least one weight vector; (25) is a hierarchical constraint that ensures that if the cumulative weight of one i higher-level criterion (k'_i) is greater than the cumulative weight of another $i + 1$ higher-level criterion (k'_{i+1}), then the sum of the cumulative weights of the lower-level criteria associated with the i criterion is greater than the sum of cumulative weights of the lower-level criteria associated with the $i + 1$ criterion; and (26) and (27) are the restrictions of non-negativity of the cumulative weights of the higher-level criteria and of non-negativity of the cumulative weights of the lower-level criteria, respectively.

In the case of the ranking problematic, the LPP presents some modifications when compared to that of the choice problematic. These modifications can be observed in the objective function and in the absence of the potential optimality constraint that doesn't belong to the set of restrictions since it is desired to rank the alternatives and not to eliminate them. Therefore, the following LPP, represented by Eq. 28 to 37, is for a ranking problematic.

$$\max_{k_{11},\ldots,k_{21},\ldots,k_{n,m_n}} \sum_{i=1}^{n}\sum_{j=1}^{m_i}(k_i.k_{ij}).v_{ij}(x_{ija}) - \sum_{i=1}^{n}\sum_{j=1}^{m_i}(k_i.k_{ij}).v_{ij}(x_{ijz}) \quad (28)$$

$s.t.$

$$\sum_{i=1}^{n} k_i = 1 \quad (29)$$

$$\sum_{j=1}^{m_i} k_{ij} = 1, \forall i, i = 1, \ldots, n \quad (30)$$

$$k_1 > k_2 > \ldots > k_{n-1} > k_n \quad (31)$$

$$k_{i1} > k_{i2} > \ldots > k_{i,j+1} > \ldots > k_{i,m_i} \quad (32)$$

$$\begin{aligned} k_{ij}.v_{ij}\left(x'_{ij}\right) &\geq k_{ij+1} - \varepsilon; j = 1, 2, \ldots, m_i \\ k_{ij}.v_{ij}\left(x''_{ij}\right) &\leq k_{ij+1} - \varepsilon; j = 1, 2, \ldots, m_i \end{aligned} \quad (33)$$

$$\sum_{j=1}^{m_i} k_i.k_{ij} = k'_i; \forall i, i = 1, \ldots, n \quad (34)$$

$$k'_i > k'_{i+1} \therefore \sum_{j=1}^{m_i} k_i.k_{ij} > \sum_{j=1}^{m_{i+1}} k_{i+1}.k_{(i+1)j} \quad (35)$$

$$k'_i \geq 0; i = 1 \; to \; n \quad (36)$$

$$k'_{ij} \geq 0; i = 1 \; to \; n; j = 1 \; to \; m_i \quad (37)$$

In this way, after each answer provided by the decision-maker, the LPP can be evaluated, whether for the choice or for the ranking problematic according to the problem being treated. Thus, the process occurs in a similar way to the FITradeoff with the difference of LPP uses. In the next section, the results and discussion of the application of the proposed new model are presented.

5 Results and Discussions

In this section, the problems analyzed are presented, Subsect. 5.1, and their results for verifying the effectiveness of the proposed mathematical model in the case of the choice problematic, Subsect. 5.2, and the ranking problematic, Subsect. 5.3.

5.1 Description of the Problems

In this subsection, three problems are presented, which were solved using the new mathematical model proposed for both choice and ranking problematic and in the following Subsects. 5.2 and 5.3 will be presented and analyzed of the results found. Some important factors to highlight are that for all problems the decision variables were the relative weights of the lower-level criteria and that the relative weights of the higher-level criteria were provided by direct definition as input by the decision-maker.

The first problem studied was adapted from Belton and Stewart (2002) and the relative weights of the higher-level criteria were $k_1 = 0{,}4$, $k_2 = 0{,}3$ and $k_3 = 0{,}3$. The second problem was adapted from Xia and Wu (2007) and the relative weights of the higher-level criteria were $k_1 = 0{,}44$ and $k_2 = 0{,}56$. In turn, the third problem was adapted from Keeney and Raifa (1993) and the relative weights of the higher-level criteria were $k_1 = 0{,}2$, $k_2 = 0{,}3$, $k_3 = 0{,}3$ and $k_4 = 0{,}2$. Figure 4, 5 and 6 represent the hierarchies and Tables 1, 2 and 3 represent the consequence matrices of the first, second and third problems, respectively.

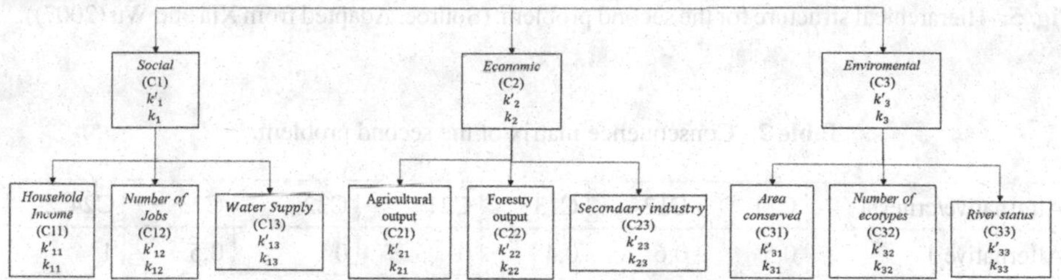

Fig. 4. Hierarchical structure for the first problem. (Source: Adapted from Belton and Stewart (2002))

5.2 Results and Discussion of the Mathematical Model – Choice Problematic

This section presents the results and discussion of the three problems presented in Subsect. 5.1 using the mathematical model for the choice problematic. In the first problem analyzed, adapted from Belton and Stewart (2002), alternative 2 was found as a result after 6 questions were answered and all families fed a uniform distribution for the relative weights of the lower-level criteria.

In the second problem analyzed, adapted from Xia and Wu (2007), alternative 4 was found as a result after 2 questions were answered and the first family presented a modal distribution while the second presented a uniform distribution for the relative weights of

Table 1. Consequence matrix of the first problem.

Alternative/criteria	C11	C12	C13	C21	C22	C23	C31	C32	C33
Alternative 1	0	0,857	0,333	0,2	0,485	1	0,5	0,5	0
Alternative 2	0,4	0,571	0,667	0,8	0,325	1	0,5	1	1
Alternative 3	0,8	0,286	1	1	0	0	0	0	0
Alternative 4	0	0	0	0	1	0,4	1	0,75	0,4
Alternative 5	1	0,286	0,286	1	0,89	0,3	0,75	0,25	0,2
Alternative 6	0,6	0,714	0,717	0,3	0,177	0,34	0,25	0,4	0,8
Alternative 7	0,2	1	1	0,6	0,257	0,8	0,4	0,75	0,6

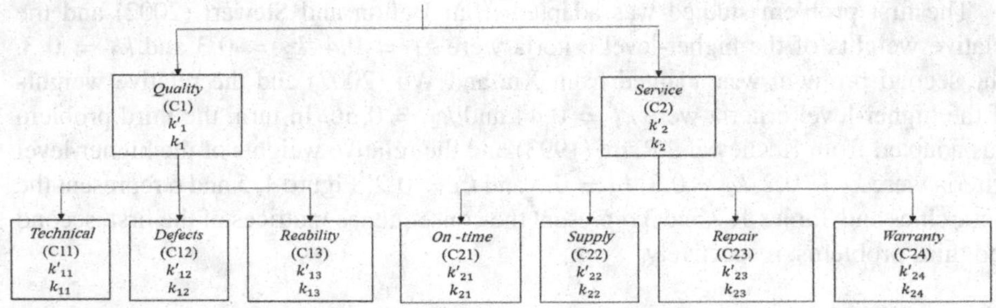

Fig. 5. Hierarchical structure for the second problem. (Source: Adapted from Xia and Wu (2007)).

Table 2. Consequence matrix of the second problem.

Alternative/criteria	C11	C12	C13	C21	C22	C23	C24
Alternative 1	0,5	0,6	0,4	0	0	0,5	1
Alternative 2	0	0	1	0,769	1	0	0
Alternative 3	0	0,2	0,8	1	0,667	0	0

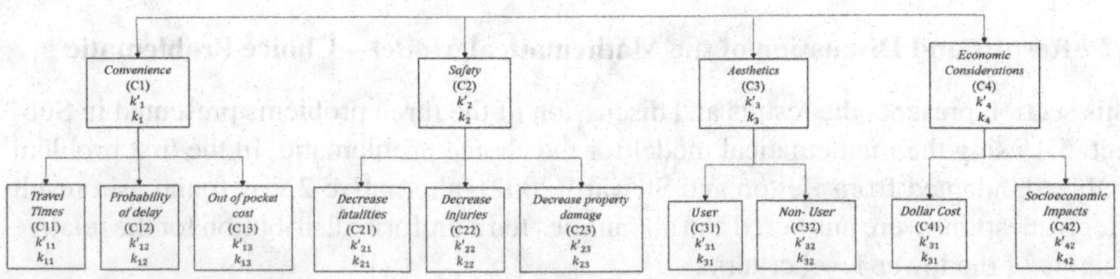

Fig. 6. Hierarchical structure for the third problem. (Source: adapted from Keeney and Raifa (1993)).

Table 3. Consequence matrix of the third problem.

Alternative/criteria	C11	C12	C13	C21	C22	C23	C31	C32	C41	C41
Alternative 1	0,922	1	0,286	0,5	0	0,493	0	0,667	0	0,556
Alternative 2	1	0	0,714	1	0,667	1	0	0	0,4	0
Alternative 3	0	0,629	0,286	0,5	0	0,32	0	1	0,6	1
Alternative 4	0,7	0,805	0	0	1	0	1	0,533	0,8	0,444
Alternative 5	0,922	0,231	1	1	0	0,432	0,5	1	1	0,778

the lower-level criteria. In turn, in the third problem analyzed, adapted from Keeney and Raifa (1993), alternative 5 was found as result after 6 questions were answered and the first family presented a modal distribution while the second and third family a uniform distribution for the relative weights of the lower-level criteria.

The graphs, in Fig. 7, illustrate the cumulative weight ranges for each lower-level criterion after the last question was answered and the unique solution of each problem studied was found. Thus, Fig. 7 represents the results of the first, second and third problems, respectively, where the lower-level criteria with similar markers color represent the lower-level criteria that belong to the same family.

Through the observation of Fig. 7, it is possible to notice that the weight range is narrow, which does not mean that the result is not good, since it is in agreement with the preferences of the decision-maker. It is important to highlight that in some real conditions and problems it is possible to find such situations. In addition, it was possible to observe that even the problems having different numbers of families, alternatives and lower-level criteria, a single solution was found in all cases showing the effectiveness of the mathematical model to solve choice problems.

Another prominent factor was the relative weights of the higher-level criteria, different in the evaluated cases, which may have been determined from the three different ways described in assumptions 1, and evidenced the fact that, for hierarchies with two levels, this assumption is sufficient to ensure that the mathematical model is effective in solving problems. In view of the results obtained and the observations made, it is possible to conclude that the mathematical model presented for the choice problematic is effective for solving problems with hierarchically structured criteria and two hierarchical levels.

5.3 Results and Discussion of the Mathematical Model–Ranking Problematic

In this section, the results and discussion for the three problems described in Sect. 5.1 are presented. The results were obtained using the mathematical model for the ranking problem. In the first problem analyzed, adapted from Belton and Stewart (2002), a partial rank of the alternatives was found as a result after 17 questions were answered and all families showed a uniform distribution for the relative weights of the lower-level criteria.

In the second problem analyzed, adapted from Xia and Wu (2007), a complete rank of the alternatives was found as a result after 6 questions were answered and the first family presented a modal distribution while the second presented a uniform distribution for the

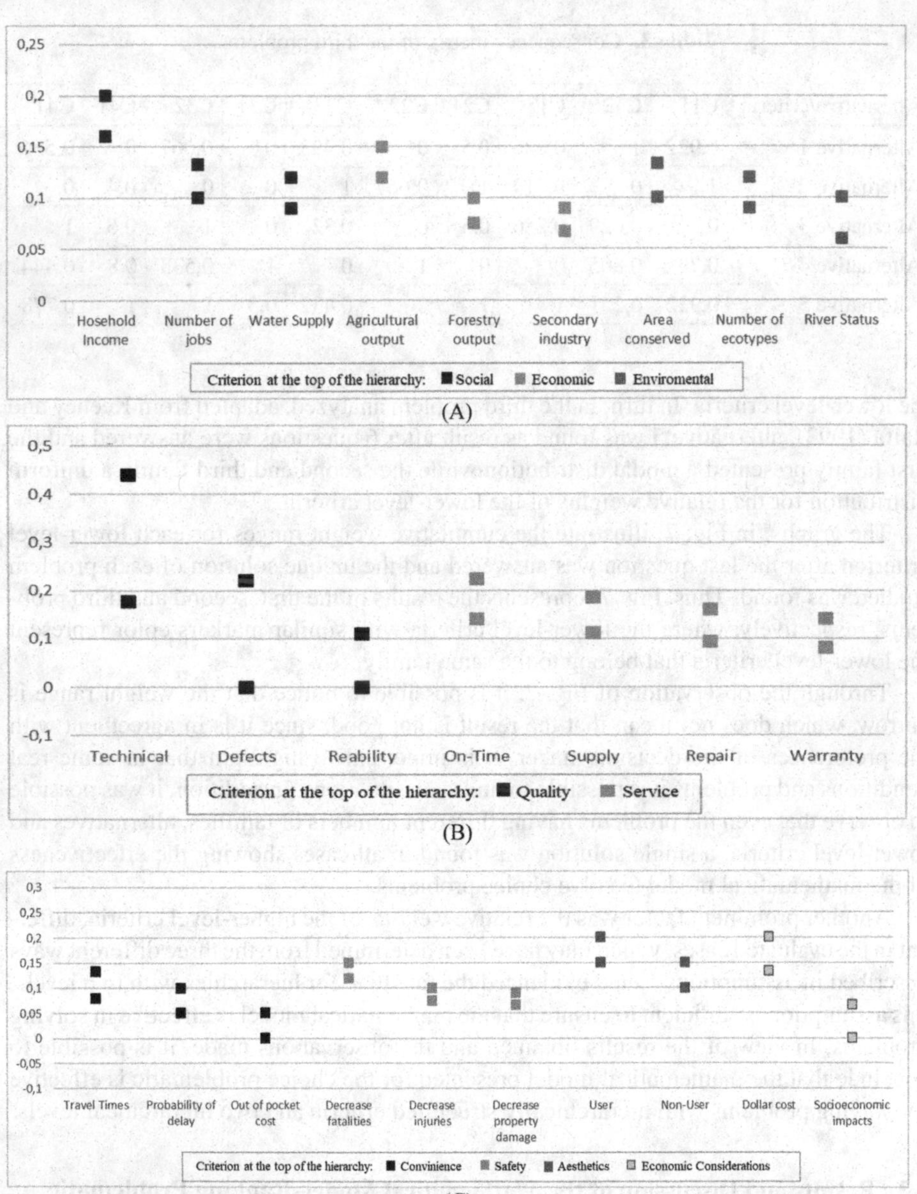

Fig. 7. Represent the ranges of each criterion for the unique solution found in the adapted problems from (A) Belton and Stewart (2002), (B) Xia and Wu (2007) and (C) Keeney and Raifa (1993).

relative weights of the lower-level criteria. In turn, in the third problem analyzed, adapted from Keeney and Raifa (1993), it was also found a complete rank of the alternatives after 11 questions were answered and the first and fourth family presented a modal distribution while the second and third family presented a uniform distribution for the relative weights of the lower-level criteria.

Figure 8 illustrates the Hass diagrams for first, second and third problems analyzed, respectively, after solving each of them. In this diagram, the directed arrows represent dominance relations and lines without arrows represent indifference between the alternatives.

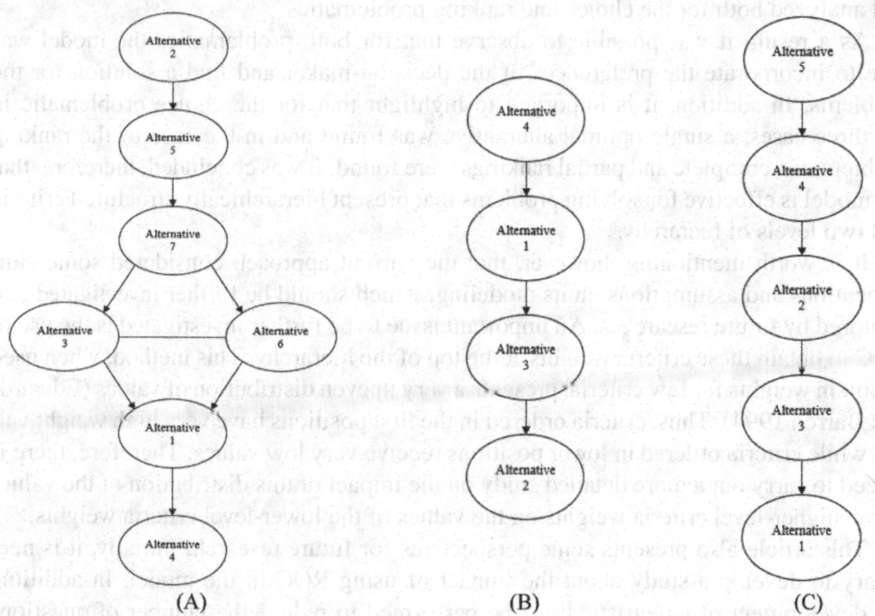

Fig. 8. Ranking visualization diagram after the last answer is provided by the decision-maker for the adapted problems from (A) Belton and Stewart (2002), (B) Xia and Wu (2007) and (C) Keeney and Raifa (1993).

Through the observation of Fig. 8, it is possible to notice that the proposed mathematical model is able to incorporate decision makers' preferences and obtain as a result both ranks with indifferent and non-indifferent alternatives. Thus, at the end of the elicitation process, it is possible to find both a complete rank or a partial rank. In addition, as in the case of the choice problematic, the model for the ranking problematic was also able to solve the problems with different characteristics such as number of families, alternatives and lower-level criteria.

Another important factor was that, as in the case of the choice problematic, the weights of the higher-level criteria defined by the decision-maker are different in the three analyzed cases. Evidencing the fact that, for hierarchies with two levels, this definition is enough to ensure that the mathematical model for the ranking problematic is also effective in solving problems. Given the above, it is possible to note that the model is capable of solving different problems that occur in the real world. Thus, being effective for solving problems in which the criteria are structured hierarchically and have two hierarchical levels.

6 Final Considerations

This article presented a new approach for dealing with hierarchical structured criteria for both choice and ranking problematics in FITradeoff. In order to verify the effectiveness of the model, three problems adapted from the literature on the value tree were presented and analyzed both for the choice and ranking problematics.

As a result, it was possible to observe that for both problematics the model was able to incorporate the preferences of the decision-maker and find a solution for the problems. In addition, it is important to highlight that for the choice problematic in the three cases, a single optimal alternative was found and in the case of the ranking problematic, complete and partial rankings were found. It was concluded, therefore, that the model is effective for solving problems that present hierarchically structured criteria and two levels of hierarchy.

It is worth mentioning, however, that the current approach considered some simplifications and assumptions on its modeling, which should be further investigated and explored by future researches. An important issue to be further investigated is the use of ROC to obtain these criteria weights at the top of the hierarchy. This method, when used to obtain weights for few criteria, presents a very uneven distribution of values (Edwards and Barron 1994). Thus, criteria ordered in the first positions have very high weight values while criteria ordered in lower positions receive very low values. Therefore, there is a need to carry out a more detailed study on the impact of this distribution of the values of the higher-level criteria weights on the values of the lower-level criteria weights.

This article also presents some perspectives for future research. Initially, it is necessary to develop a study about the impact of using ROC in the model. In addition, the development of a heuristic must be performed to reduce the number of questions answered by the decision-maker, aiming to reduce its cognitive effort. Lastly, another field of research would be the implementation of this model for problems with more than two hierarchical levels, since at first this still an existing limitation.

Acknowledgments. This work had support from the Brazilian Research Council (CNPq) and Coordination for the Improvement of Higher Education Personnel (CAPES).

References

Belton, V., Stewart, T.: Multiple Criteria Decision Analysis: An Integrated Approach. Springer, Boston (2002). https://doi.org/10.1007/978-1-4615-1495-4

Belton, V., Goodwin, P.: Remarks on the application of the analytic hierarchy process to judgmental forecasting. Int. J. Forecast. **12**, 155–161 (1996). https://doi.org/10.1016/0169-2070(95)00643-5

Bouyssou, D., Jacquet-Lagrèze, E., Perny, P., Slowiński, R., Vanderpooten, D., Vincke, P.: Aiding Decisions with Multiple Criteria. Springer, New York (2002). https://doi.org/10.1007/978-1-4615-0843-4

Carrillo, P.A.A., Roselli, L.R.P., Frej, E.A., de Almeida, A.T.: Selecting an agricultural technology package based on the flexible and interactive tradeoff method. Ann. Oper. Res. (2018). https://doi.org/10.1007/s10479-018-3020-y

De Almeida, A.T., Almeida, J.A., Costa, A.P.C.S., Almeida-Filho, A.T.: A new method for elicitation of criteria weights in additive models: flexible and interactive tradeoff. Eur. J. Oper. Res. **250**, 179–191 (2016). https://doi.org/10.1016/j.ejor.2015.08.058

de Almeida, A.T., Frej, E.A., Roselli, L.R.P.: Combining holistic and decomposition paradigms in preference modeling with the flexibility of FITradeoff. CEJOR **29**(1), 7–47 (2021). https://doi.org/10.1007/s10100-020-00728-z

Dell'ovo, M., Oppio, A., Capolongo, S.: Sistema de Apoio à Decisão para a Localização de Estabelecimentos de Saúde: ferramenta de avaliação sithealth. Springer, Milão (2020)

Edwards, W., Barron, F.H.: SMARTS and SMARTER: improved simple methods for multiattribute utility measurement. Organ. Behav. Hum. Decis. Process. **60**, 306–325 (1994). https://doi.org/10.1006/obhd.1994.1087

Frej, E.A., Roselli, L.R.P., Almeida, J.A., de Almeida, A.T.: A multicriteria decision model for supplier selection in a food industry based on FITradeoff Method. Hindawi (2017). https://doi.org/10.1155/2017/4541914

Frej, E.A., de Almeida, A.T., Costa, A.P.C.S.: Using data visualization for ranking alternatives with partial information and interactive tradeoff elicitation. Oper. Res. Int. J. **19**(4), 909–931 (2019). https://doi.org/10.1007/s12351-018-00444-2

Frej, E.A., Ekel, P., de Almeida, A.T.: A benefit-to-cost ratio based approach for portfolio selection under multiple criteria with incomplete preference information. Inf. Sci. **545**, 487–498 (2021). https://doi.org/10.1016/j.ins.2020.08.119

Fossile, D.K., Frej, E.A., Costa, S.E.G., Lima, E.P., de Almeida, A.T.: Selecting the most viable renewable energy source for Brazilian ports using the FITradeoff method. J. Clean. Prod. **260** (2020). https://doi.org/10.1016/j.jclepro.2020.121107

Kajanus, M., Kangas, J., Kurttila, M.: The use of value focused thinking and the A'WOT hybrid method in tourism management. Tour. Manag. **25**, 499–506 (2004). https://doi.org/10.1016/S0261-5177(03)00120-1

Keeney, R.L.: Value-Focused Thinking: A Path to Creative Decision Making. Harvard University Press, Cambridge (1992)

Keeney, R.L.: Value-focused thinking: Identifying decision opportunities and creating alternatives. Eur. J. Oper. Res. **92**, 537–549 (1996). https://doi.org/10.1016/0377-2217(96)00004-5

Keeney, R.L., Raiffa, H.: Decisions with Multiple Objectives. Cambridge University Press, Cambridge (1993)

Keeney, R.L., Raiffa, H.: Decision Analysis with Multiple Conflicting Objectives. Cambridge University Press, New York (1976)

Louvieris, P., Gregoriades, A., Garn, W.: Assessing critical success factors for military decision support. Expert Syst. Appl. **37**, 8229–8241 (2010). https://doi.org/10.1016/j.eswa.2010.05.062

Pergher, I., Frej, E.A., Roselli, L.R.P., De Almeida, A.T.: Integrating simulation and FITradeoff method for scheduling rules selection in job-shop production systems. Int. J. Prod. Econ. **227** (2020). https://doi.org/10.1016/j.ijpe.2020.107669

Poyhonen, M., Vrolijk, H., Hamalainen, R.P.: Behavioral and procedural consequences of structural variation in value trees. Eur. J. Oper. Res. **134**, 216–227 (2001). https://doi.org/10.1016/S0377-2217(00)00255-1

Poyhonen, M., Hamalainen, R.P.: Notes on the weighting biases in value trees. J. Behav. Decis. Mak. **11**, 139–150 (1998). https://doi.org/10.1002/(SICI)1099-0771(199806)11:2%3c139::AID-BDM293%3e3.0.CO;2-M

Rodrigues, L.V.S., Casado, R.S.G.R., Carvalho, E.N., Silva, M.M.: Using FITradeoff in a ranking problem for supplier selection under TBL performance evaluation: an application in the textile sector. Production **30** (2020). https://doi.org/10.1590/0103-6513.20190032

Roselli, L.R.P., Frej, E.A., de Almeida, A.T.: Neuroscience experiment for graphical visualization in the FITradeoff decision support system. In: Chen, Y., Kersten, G., Vetschera, R., Xu, H. (eds.) GDN 2018. LNBIP, vol. 315, pp. 56–69. Springer, Cham (2018). https://doi.org/10.1007/978-3-319-92874-6_5

Saaty, T.L.: The Analytic Hierarchy Process. McGraw-Hill, New York (1980)

Salo, A.A., Hamalainen, R.P.: Preference assessment by imprecise ratio statements. Oper. Res. Int. J. **40**, 1053–1061 (1992). https://doi.org/10.1287/opre.40.6.1053

Salo, A.A., Hamalainen, R.P.: Preference ratios in multiattribute evaluation (PRIME)—elicitation and decision procedures under incomplete information. IEEE Trans. Syst. Man Cybern. Part A Syst. Hum. **31**, 533–545 (2001). https://doi.org/10.1109/3468.983411

Salo, A., Punkka, A.: Rank inclusion in criteria hierarchies. Eur. J. Oper. Res. **163**, 338–356 (2005). https://doi.org/10.1016/j.ejor.2003.10.014

Von Winterfeldt, D., Edwards, W.: Decision Analysis and Behavioral Research. Cambridge University Press, New York (1986)

Xia, W., Wu, Z.: Supplier selection with multiple criteria in volume discount environments. Omega **35**(5), 494–504 (2007). https://doi.org/10.1016/j.omega.2005.09.002

Weber, M., Borcherding, K.: Behavioral influences on weight judgments in multiattribute decision making. Eur. J. Oper. Res. **67**, 1–12 (1993). https://doi.org/10.1016/0377-2217(93)90318-H

A User Interface for Consistent AHP Pairwise Comparisons

Andrés Cimadamore[1], Alejandro Fernandez[1], Chenhui Ye[2], Pascale Zaraté[2](✉), and Daouda Kamissoko[3]

[1] LIFIA, CICPBA, FI, Universidad Nacional de La Plata, Argentina Calle 20 y 120, (1900) La Plata, Argentina
{andres.cimadamore,alejandro.fernandez}@lifia.info.unlp.edu.ar
[2] IRIT, Toulouse Université, 2 rue du Doyen Gabriel Marty, 31042 Toulouse Cedex 9, France
Pascale.Zarate@irit.fr
[3] Ecole Des Mines d'Albi, Albi, France
Dadouda.Kamissoko@mines-albi.fr
http://www.lifia.info.unlp.edu.ar/,
https://www.irit.fr/~Pascale.Zarate/

Abstract. Decision Makers generally reason on several criteria, aiming to obtain a total consistency or partial order of several alternatives. MultiCriteria analysis is based on the assumption that such ordering exists. Decision Makers are supported by several kinds of approaches or tools. One approach consists in comparing the criteria two by two, i.e. pairwise comparison, to find the relative importance of each criterion. This relative importance, called weight of criteria, is used to find the final order of alternatives. One methodology, developed by Saaty, called Analytical Hierarchical Process (AHP), is based on this principle of pairwise comparison. Having the weights of criteria, the decision makers have then to compare the alternatives two by two for each criterion. Pairwise comparisons are simple to use; however, as the number of items to compare increases, so do the effort of conducting all comparisons and the probability of introducing inconsistencies. In this article we present an innovative approach to conduct pairwise comparisons based on a UI widget that resembles an interactive data plot. It uses the transitivity property of a consistent comparison matrix to infer comparisons. Our hypothesis is that this new approach is more efficient (as it reduces the number of actions the user must conduct to compare all items), more effective (as it limits the sources of inconsistencies), and yields better user satisfaction. We conducted a controlled experiment involving 50 participants. We observed that the proposed widget reduces the effort of making pairwise comparisons, improves the consistency of the comparisons, and leads to a better user experience.

Keywords: Pairwise comparisons · Consistency · Transitivity · UI design · Usability · AHP

A. T. de Almeida and D. C. Morais (Eds.): INSID 2021, LNBIP 435, pp. 119–134, 2021.
https://doi.org/10.1007/978-3-030-91768-5_8

1 Introduction

The AHP method is an analytical approach for supporting decision making following a multi-criteria approach [1]. It has been used in several areas, such as transport planning, rationing of energy, risk management projects, benchmarking of logistics operations, management of quality of services in hospitals, operations management, allocation resources for product portfolio management. It was developed by Thomas Saaty in 1970 and allows the decomposition of a complex problem in a hierarchical system. Alternatives defined by the decision maker provide their relative priorities thanks to a pairwise comparison. Then a synthesis allows decision makers to easily understand what would be the best choice. Classification is performed at several levels which are associated with different criteria. Thus, it is possible to determine the most appropriate alternative, depending on the priority given to each used criteria. Pairwise comparisons (PCs) are a central feature of AHP.

In this article we present an approach to conduct PCs that is easy to use, intuitive, reduces the number of required comparisons, and yields consistent and complete comparison matrices. A visual 2D representation of the comparable items is used to express relative preferences among items. The transitivity property of the AHP matrix is used to infer preferences thus reducing the number of required comparisons. As a result, the method yields more consistent matrices regardless of the number of alternatives considered.

Next, we motivate our work by presenting an overview of key concepts regarding consistency and transitivity in AHP comparisons, and by discussing the role of visualizations. Then, we present our approach based on an innovative pairwise comparison widget. Following, we present the methodology used to evaluate the approach and the results we obtained. To conclude, we offer conclusions and discuss future work.

2 Background: Pairwise Comparisons in AHP

Following the construstruction of the hierarchical model with various levels of criteria and one one level of alternatives, PCs are carried out at each level. Different scales can be used to compare items [2]. In this work we focus on the original scale proposed by Saaty, using integer values in [1, 9], and their reciprocals. The decision makers' judgments are kept in a matrix model called the Judgments Matrix. The main objective is to compare the relative importance of all elements belonging to the same level.

Transitivity in multiple criteria decision making is also called ordinal consistency [3]. If a decision maker prefers alternative $x1$ to alternative $x2$ and $x2$ to $x3$, then transitivity requires that he/she also prefers $x1$ to $x3$, as otherwise, cycles would exist in the preferences. Tversky [4] considered transitivity to be the cornerstone of normative decision theory. Preference transitivity is a basic principle in most major rational, descriptive decision models [5].

Benitez et al. [6] propose a method to achieve consistency in AHP through optimisation. This method has the major advantage of depending on just the decision variables – the number of compared elements – and so is less computationally expensive than other optimisation methods, and can be easily implemented in virtually any existing computer environment.

Decision support software packages such as Super Decisions [7] and Expert Choice [8] offer alternative modes to elicit user preferences as PCs. A frequent strategy to elicit comparisons is to present them in a matrix. Each value in the cell compares the item represented by the row, to the item represented by the column. Entering a value in a cell, automatically updates the value if its inverse. This method requires users to get accustomed to the direction of the comparisons, and the interpretation of the values (which are both integers and fractions). Super Decisions improves the matrix view by removing the values in the diagonal, keeping only one value for each pair (i.e., removing the inverse comparison), and introducing an arrow that indicates the direction of the comparison (see Fig. 1).

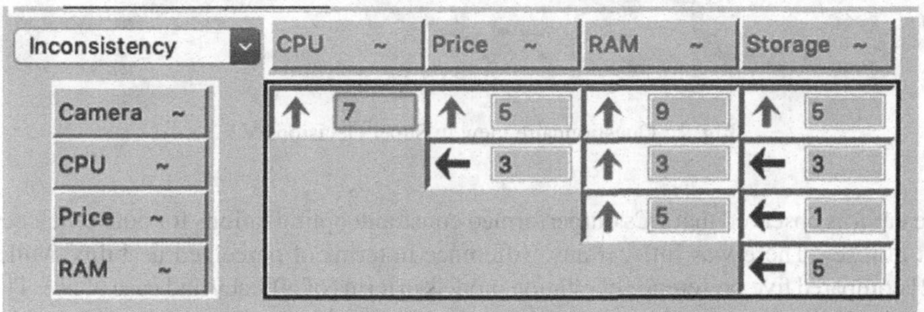

Fig. 1. Matrix view in Super Decisions V3.2

The questionnaire view is another common presentation to elicit PCs. Each row represents one pairwise comparison, with the items to compare on each side. The user must place a mark closest to the item that is considered more important (or preferred). Super Decisions uses radio buttons as markers in its questionnaire view (see Fig. 2), while PriEst [9] uses sliders (in PriEst, this view is called equalizer). Placing the marker in the middle indicates that items are equally important. The number of positions between items normally reflect values from 2 to 9 in each direction, plus 1 in the middle position.

Both, the matrix presentation and the questionnaire presentation, offer a holistic view of all comparisons. In addition to these holistic presentations, some tools offer visual means to manipulate individual comparisons, for example in the form of an interactive bar or pie chart.

All the aforementioned strategies consider that each pairwise comparison is independent from the rest of them. Independence among comparisons directly correlates to consistency; the more liberty (and the less scaffolding) users have to independently compare items the more likely they are to introduce inconsistency. Super Decisions and PriEsT offer help to identify inconsistency. In addition, PriEsT offers visual aids to observe transitivity.

How to present and to elicit PCs is one of the challenges faced by decision support tools designers; modelling preferences is almost as important as the modelling of the logical structure of the problem [10]. Abel et al. [11] compared the usability of two contrasting approaches to elicit decision priorities namely, PCs and constrained optimizations. Their work focuses on performance and usability as perceived by the user.

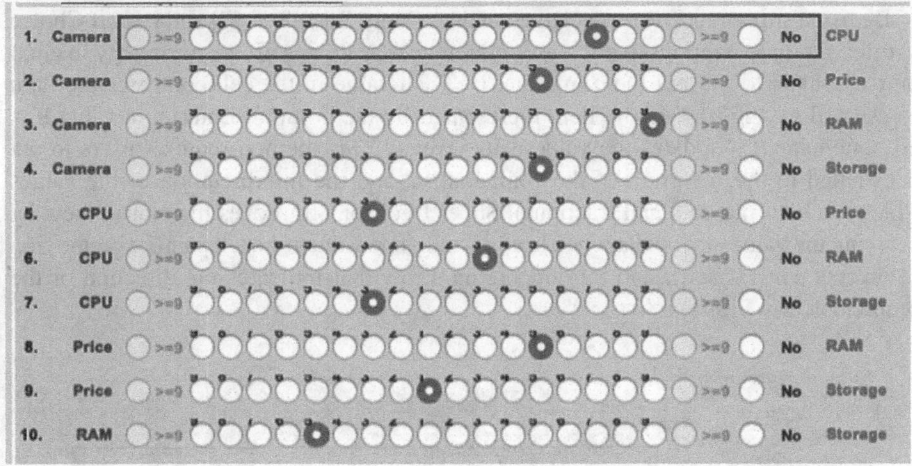

Fig. 2. Questionnaire view in Super Decisions V3.2

The authors observed that PCs outperformed constraint optimizations for both efficiency and efficacy. There was little, if any, difference in terms of perceived usability. Millet [12] compared five preference elicitation models in terms of efficacy and ease of use. The results of Millet's research supports the motivation of this work to explore alternative graphical modes to elicit preferences.

Perfectly stating pairwise preferences is seldomly possible for a decision maker. Many factors (such as the number of possible transitive steps) can cause the introduction of inconsistencies in a decision matrix that results from pairwise comparison. Computing the consistency of a decision matrix is a means to assess the decision maker's understanding and experience in a field (which can help value decisions). There are many methods to compute the consistency of a decision matrix [13]. In this work we assess the quality of a given comparison matrix by means of the Consistency Ratio (CR) as proposed by Saaty [14]. It was introduced by Saaty [14] in order to check decision-makers preferences consistency in the AHP methodology. This CR was then analysed by several authors and is one of the main ways to check this consistency. It is the reason why we decided to use it in our study.

Computing the CR is a two step process. First, the Consistency Index (CI) of a matrix A of size n is computed according to Eq. 1, where λmax is the maximum eigenvalue of the matrix. Then, the CR is the ratio between the CI and a real number called the Random Index (Eq. 2). The random index for a matrix of size n (RIn) is an estimation of the average CI obtained from a large set of randomly generated matrices of size n [15].

$$CI(A) = \frac{\lambda max - n}{n - 1} \tag{1}$$

$$CR(A) = \frac{CI(A)}{RIn} \tag{2}$$

3 Transitive Spacial Comparisons

Our approach to support PCs while maintaining consistency builds on two pillars. Firstly, it proposes a new visual tool (a User Interface widget) to express relative preferences. Secondly, all PCs are updated on every preference update using the transitivity property of an (assumed) consistent AHP matrix. The design of the widget conveys the transitivity of comparisons.

The proposed widget is depicted in Fig. 3. It resembles a 2D, continuous data plot. The vertical axis is labeled with the expressions that are normally given to the values in Saaty's scale. All items to compare are placed on the horizontal axis. The plot line includes a handle (a small circle) for the value corresponding to each item. The first handle (in this case, Price) is "anchored" to the middle value (representing 1, or equally important). All other handles can be moved upwards or downwards.

At first, all handles are anchored at the middle position indicating that they are equally important. The user moves handles to indicate how a given item compares to the anchored one. For example, Fig. 3 shows that the handle for RAM has been moved upwards to indicate that RAM is very strongly better than Price (the anchored item). Moreover, moving a handle to express how the item compares to the anchored one, also indicates how it compares to all other items. That is, moving the handle for RAM expresses how it compares to the anchored item (Price), but also to all other items.

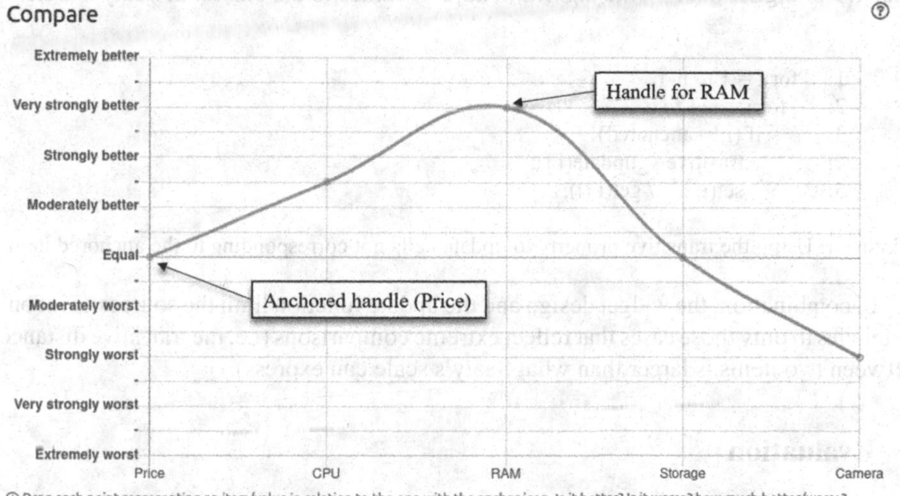

Fig. 3. UI widget to present and elicit PCs

The widget offers a complete picture of how items compare to one another; however, labels in the vertical axis are expressed only in relation to the anchored item. Double clicking on a handle (the small circle on the plot), anchors the criterion for that handle to the left, updating the position of all other handles to reflect the change. This feature lets users explore the comparison space from the perspective of each item.

The visual representation of the comparisons helps maintain transitivity. By plotting all preferences in a 2D space, they inherit the transitivity properties of the comparison function in real numbers. Moving one handle simultaneously expresses how the item compares to the anchored one, and to all other items. This feature of the widget reduces the number of actions (e.g., clicks) that the user must make to compare all items.

PCs are stored in a matrix using Saaty's scale; rows and columns represent the handles (i.e., the items that handles stand for). The widget is divided in the middle by a line labeled "equally important". In addition, each half is divided by eight lighter lines (that correspond to the labels in the vertical axis).

When the user moves a handle *h*, the widget computes which line it is closest to. Then it computes the vertical distance d_i (rounded to the next integer) from that line, to the horizontal axis. The distance d_i is used to update the value in the cell that corresponds to the comparison between *h* and the anchored item. If *h* is above the horizontal axis, the cell takes the value $1/(d_i+1)$; otherwise, it takes $(d_i + 1)$.

As described previously, weights in an AHP matrix are consistent if they are transitive. That is, $a_{ik} = a_{ij}a_{jk}$ for all *i*, *j*, and *k*. The widget forces the transitivity property, using it to compute all cell values for rows different from that of the anchored item. Listing 1 outlines the algorithm used to transitively update cells. It iterates only over the cells that are above the diagonal and not in the row that corresponds to the anchored item. For each cell, it sets the expected value (according to the transitivity formula), and it sets the value of its inverse. To deal with rounding errors and extreme values (smaller than 1/9 or higher than 9), the algorithm adjusts values to the closest in Saaty's scale.

```
for i = 1 to n-1
  for j = i+1 to n
    if (i != anchored)
      trasitively_update(i,j);
      set(j, i, 1 / get(i,j));
```

Listing 1: Using the transitive property to update cells not corresponding to the anchored item

In combination, the widget design and the update function limit the sources of inconsistencies to only those cases that reflect extreme comparisons (i.e. the transitive distance between two items is larger than what Saaty's scale can express).

4 Evaluation

We argue that the proposed approach represents an improvement in terms of usability and consistency of the resulting comparison matrix. To formally demonstrate such a claim, we conducted a controlled experiment. The general design of the experiment involves comparing the proposed widget presented in Sect. 3 to the widespread questionnaire widget depicted in Sect. 2, Fig. 2. The experiment aims to answer the following questions:

- Efficiency: Does the new widget reduce the effort of conducting pairwise comparisons?

- Efficacy: Does the new widget improve the consistency ratio of the resulting comparison matrix?
- Ease of use: Is the new widget easier to use than the questionnaire widget?
- Is there a correlation between the use of one widget or the other and the perceived validity of the resulting ranking?

The experiment was designed as a sequence of tasks to be completed, and surveys to complete. Following we describe the experiment protocol in more detail.

4.1 Participants

The experiment involved 45 students from a computer and information system Master in Toulouse as the experimental subjects. All 45 students have received at least one year of computer training, and master basic decision-making knowledge and clear logic. In addition, 5 researchers, holding PhDs in computer science from the Toulouse-IRIT laboratory took part. The researchers are not related to the project. All participants in the experiment have a basic understanding and basic knowledge in the field of decision support.

4.2 Evaluation Protocol

The experiment was carried out individually, one participant at a time. A Within-Subjects design was used for this study, meaning that each subject used both pairwise comparison widgets.

Figure 4 provides an overview of the experiment protocol. First, participants receive instructions related to the tasks they will be asked to perform. Instructions include a brief introduction to AHP and the widget that will be used. Special care was taken to avoid guiding the participants to any expected outcome. Then, each participant conducts comparisons in two different scenarios. Each scenario includes two pairwise comparisons, one of them on a tangible criteria and the other one on an intangible criteria. In one scenario participants compare travel destinations in terms of price (tangible) and attractions (intangible). In the other scenario, participants compare apartments in terms of price (tangible) and characteristics (intangible). Both pairwise comparisons in each scenario must be conducted with one of the two widgets. Both widgets must be used (one for each scenario).

To simplify the experiment, participants are not asked to compare criteria, only alternatives. Criteria are defined to be equally important. After participants finish both pairwise comparisons in one scenario, the resulting ranking of alternatives is presented, and discussed. If the resulting ranking does not match the one expected by the participant, the participant is asked to write down an alternative one.

After completing each scenario, participants answer a satisfaction survey. Finally, after completing both scenarios, participants answer a comparative survey. This experiment design results in four different combinations of scenarios and widgets (see Table 1).

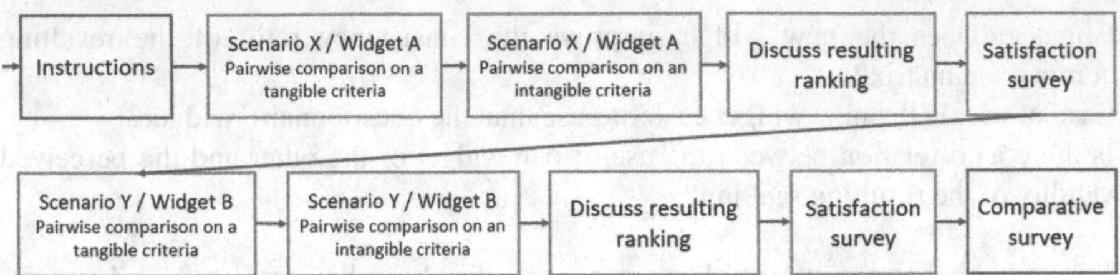

Fig. 4. Overview of the experiment session

Table 1. Four possible combinations of scenarios and widgets to avoid learning effect and task bias

Scenario X	Widget A (Tangible criteria)	Widget A (Intangible criteria)	Scenario Y	Widget B (Tangible criteria)	Widget B (Intangible criteria)
Choosing a travel destination	Questionnaire widget (on price)	Questionnaire widget (on attractions)	Choosing an apartment	2D Plot widget (on price)	2D Plot widget (on characteristics)
Choosing a travel destination	2D Plot widget (on price)	2D Plot widget (on attractions)	Choosing an apartment	Questionnaire widget (on price)	Questionnaire widget (on characteristics)
Choosing an apartment	Questionnaire widget (on price)	Questionnaire widget (on characteristics)	Choosing a travel destination	2D Plot widget (on price)	2D Plot widget (on attractions)
Choosing an apartment	2D Plot widget (on price)	2D Plot widget (on characteristics)	Choosing a travel destination	Questionnaire widget (on price)	Questionnaire widget (on attractions)

4.3 Tools and Data Collection

Both comparison widgets were deployed as part of an ad-hoc web application. The Plot 2D widget was implemented as depicted in Fig. 3 (in the previous Section). The questionnaire widget was implemented as depicted in Fig. 4 below.

It collects the start time and end time for each scenario, and for each pairwise comparison. It also records the number of mouse clicks to complete each pairwise comparison. This information is used to compute efficiency. In this experiment, efficacy is defined in terms of the CR of each comparison matrix. The application transparently generates the comparison matrix that corresponds to each pairwise comparison. Then, it computes the CR using the procedures presented in Sect. 2.1. After each subject completed the experiment, the following data was exported from the application and collected in a spreadsheet for later analysis:

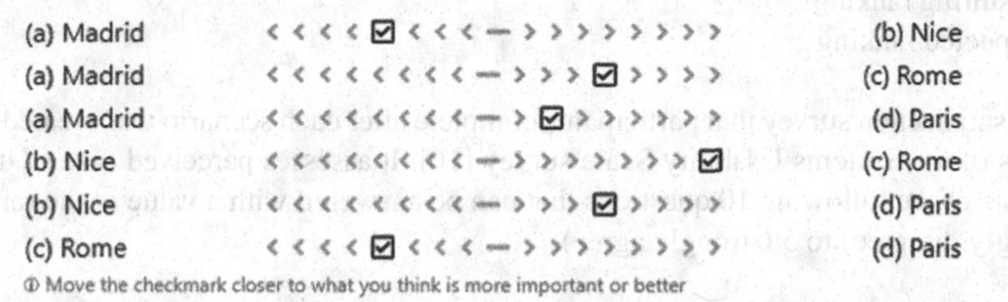

Fig. 5. Questionnaire widget as implemented in the web application used for the experiment

- Title of scenario X
  - Name of 1st criteria (tangible)
    - Name of the widget A
    - Time (ms) to complete this PC
    - Clicks to complete this PC
    - CR of this PC
  - Name of 2nd criteria (intangible)
    - Name of the widget A
    - Time (ms) to complete this PC
    - Clicks to complete this PC
    - CR of this PC
  - Resulting ranking
  - Expected ranking
- Title of scenario Y
  - Name of 1st criteria (tangible)
    - Name of the widget B
    - Time (ms) to complete this PC
    - Clicks to complete this PC
    - CR of this PC
  - Name of 2nd criteria (intangible)
    - Name of the widget B
    - Time (ms) to complete this PC
    - Clicks to complete this PC
    - CR of this PC

- Resulting ranking
- Expected ranking

The satisfaction survey that participants complete after each scenario was created on the basis of the Systems Usability Scale survey [16]. It assesses perceived ease of use. It consists of the following 10 questions that can be answered with a value in the range 1 (strongly disagree) to 5 (strongly agree).

1. I think I'll use "this comparison widget" frequently.
2. I find "this comparison widget" unnecessarily complex.
3. I think "this comparison widget" is easy to use.
4. I think I will need the help of a technician to be able to use "this comparison widget".
5. I found that the various functions of "this comparison widget" were integrated well.
6. I think there is too much inconsistency in "this comparison widget".
7. I imagine most people would be able to learn to use "this comparison widget" very quickly.
8. I found "this comparison widget" very cumbersome to use.
9. I felt very confident using "this comparison widget."
10. I need to learn a lot of things before I can use "this comparison widget".

The final comparative survey includes only the following two questions:

1. Which tool allowed you to model more precisely the relative importance of each alternative?
2. Which Tool is easiest to use?

4.4 Results

After 35 days of experiment, a total of 50 persons participated in the experiment. After data screening and cleaning, there were 13 sets of data with missing data and incomplete information. Therefore, we conducted data analysis and mining for the remaining 37 sets of data. Following, we present results one research question at a time.

- Efficiency: Does the new widget reduce the effort of conducting pairwise comparisons?

To compare both tools in terms of efficiency, we considered the time it took to complete each PC, and the number of clicks it required. Table 2 presents the results. Some samples were discarded as they presented invalid data, such as a very short time in the order of a few seconds, which suggested that the participant did not complete the task. Comparing the means for the numbers of clicks of both samples, yields a difference of 5.98 clicks less for the 2D Plot. Given the sample size and the standard deviation of the sample, this indicates that the 2D Plot requires significantly less effort with significance value (P-value) of 0.0001. In terms of time, the Plot 2D did not show a statistically significant improvement over the questionnaire widget.

Table 2. Comparison of time and number clicks to complete PC

	Clicks (Mean)	Clicks (Std)	Time (Mean)	Time (Std)	Data points
Questionnaire widget	16.76	4.82	352.135	185.756	37
2D Plot Widget	10.78	3.61	269.324	173.828	37

- Efficacy: Does the new widget improve the consistency ratio of the resulting comparison matrix?

To answer this question, we computed the CR of the resulting comparison matrix after each pairwise comparison. In each case we computed the pass rate (i.e., whether the CR was lower than 0.1).

Using tool A and tool B to make decisions in the same scenario, the ratio of CR in the consistency test results is less than 0.1. For the AHP model, consistency is a very important issue, and it is also one of the limitations of the AHP model. Generally, the AHP model cannot arrange the order of more than ten criteria or more than ten alternatives, because once the ten factors are exceeded, it is difficult to avoid inconsistencies. Especially for untrained decision makers, it is difficult to maintain logic and consistency when comparing multiple factors. A major advantage of the new approach is that it can avoid the problem of inconsistency in the user's ranking.

According to the rules of the AHP model, the CR needs to be less than 0.1. If the CR is less than 0.1, then we consider the consistency test to be passed, if it is greater than or equal to 0.1, then it is deemed that the consistency test fails. Each Scenario has two Criteria, one of them of tangible nature, and the other one of intangible nature. Table 3 presents the results regardless of the type of criteria, and Table 5 and 6 discriminate between tangible and intangible criteria. In all cases, the 2D Plot widget leads to better consistency.

Table 3. CR regardless of the type of criteria

	CR (Mean)	CR (Std)	Pass rate	Data points
Questionnaire widget	0.21494	0.18189	28.38%	74
2D Plot Widget	0.03669	0.06868	90.54%	74

- Is there a correlation between the use of one widget or the other and the perceived validity of the resulting ranking?

At the end of each scenario (i.e., after completing the two required PCs) the application presented a final ranking of alternatives. The participant had to indicate if the presented ranking was the expected one or not. If not, the participant indicated what

Table 4. CR for tangible criteria

	CR (Mean)	CR (Std)	Pass rate	Data points
Questionnaire widget	0.18681	0.1621	32.43%	37
2D Plot Widget	0.02408	0.0495	91.89%	37

Table 5. CR for intangible criteria

	CR (Mean)	CR (Std)	Pass rate	Data points
Questionnaire widget	0.24307	0.19537	24.32%	37
2D Plot Widget	0.04931	0.08163	89.19%	37

was the expected ranking. Figure 7 reports the count of matches and misses by widget. It was not a surprise to discover that misses largely outcount matches for both widgets. Whether or not the final ranking matches the user's expectations depends on multiple factors. Firstly, forcing both criteria to be equally important possibly contradicts the model the participants would have constructed. Second, if two alternatives obtain the same rank the tool randomly decides which one to rank first.

The focus of this experiment is on the effect of different widgets used in the decision making process (in PC in particular), not the whole AHP model and process. The key question is whether both widgets integrated similarly into the whole model. To gain further insight into the relation between the widget and the final ranking, we computed the Levenshtein distance between the observed and the expected ranking. Figure 7 reports the results for all misses. It can be observed that both widgets obtained a similar distribution. This result indicates that there is no significant correlation between the widget, and between the expected and observed ranking.

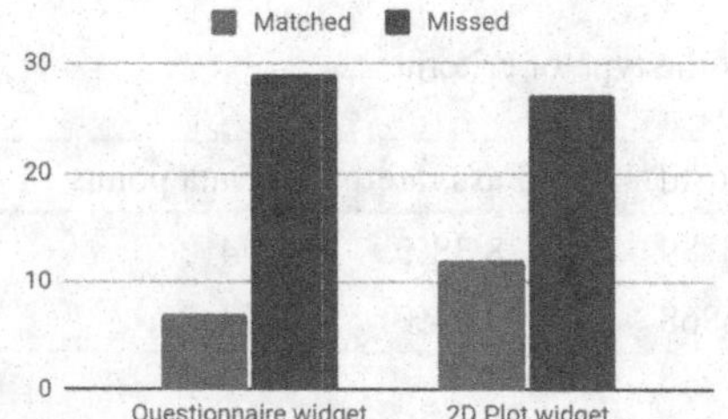

Fig. 6. Correspondence of expected rank vs. observed rank, by widget

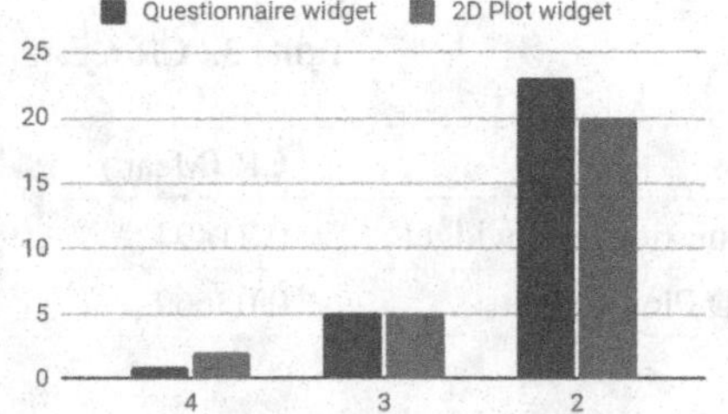

Fig. 7. Levenshtein distance between expected and observed rank, by widget

- Ease of use: Is the new widget easier to use than the questionnaire widget?

The satisfaction (perceived ease of use) survey was completed for both widgets by 30 participants. Responses were used to produce a score in the range 0–100 as suggested by the SUS method. Results were interpreted as relative to the observations made by Sauro [17]. According to the author, who conducted a large number of usability studies and compared the results, any score value under 68 is considered to be below average. Scores above 68 can be grouped in three buckets, each representing the top 30% (grade C, for scores between 68 and 74), top 20% (grade B, for scores between 74 and 83.1), and top 10% (grade A). Figure 8 presents the scores obtained by both widgets, grouping them in the above average buckets. The Plot 2D widget was perceived above average for more participants, and with higher grades in general.

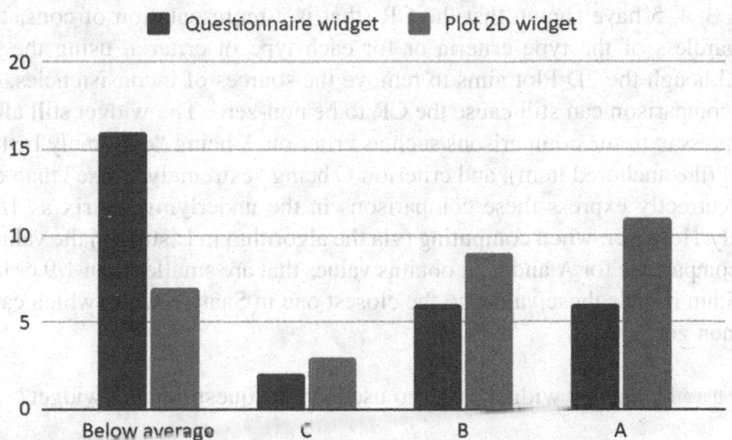

Fig. 8. SUS Scores for both widgets, for 30 responses.

The comparative survey that users completed after finishing the experiment confirms that the users perception favored the 2D Plot widget. As depicted in Fig. 9, the 2D Plot widget allowed users to model more precisely the relative importance of each alternative, and was easier to use.

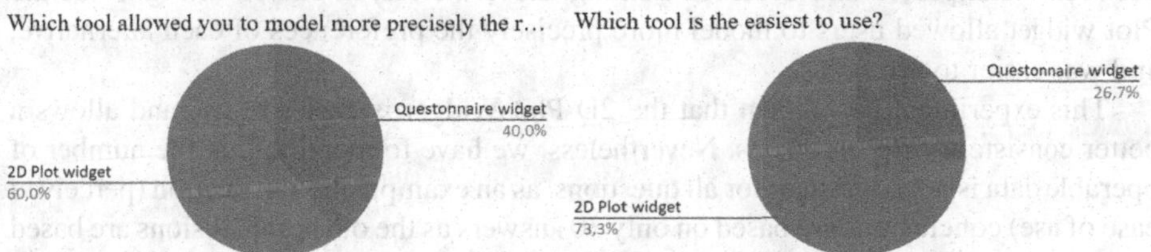

Fig. 9. Results of the comparative survey

4.5 Discussion

Our aim was to show that the proposed approach represents an improvement in terms of usability and consistency of the resulting comparison matrix. The experiment was designed to answer the following questions:

- Efficiency: Does the new widget reduce the effort of conducting pairwise comparisons?

The experiment has shown that the number of clicks and the time are clearly reduced using the 2D Plot Widget (see Table 2).

- Efficacy: Does the new widget improve the consistency ratio of the resulting comparison matrix?

Tables 3, 4, 5 have shown that the CR (that is a representation of consistency) is better, regardless of the type criteria or for each type of criteria, using the 2D Plot Widget. Although the 2D Plot aims to remove the sources of inconsistencies, extreme transitive comparison can still cause the CR to be non-zero. The widget still allows the user to express extreme comparisons such as criterion A being "extremely better" than criterion B (the anchored item), and criterion C being "extremely worse" than criterion B. It can correctly express these comparisons in the underlying matrix as 1/9 and 9 respectively. However, when computing (via the algorithm in Listing 1) the value for the pairwise comparison for A and C, it obtains values that are smaller than 1/9 or higher 9. The algorithm rounds these values to the closest one in Saaty's scale, which causes the CR to be non-zero.

- Ease of use: Is the new widget easier to use than the questionnaire widget?

The Plot 2D widget was perceived above average for more participants, and with higher grades in general.

- Is there a correlation between the use of one widget or the other and the perceived validity of the resulting ranking?

Figures 6, 7 indicate that there is no significant correlation between the widget, and between the expected and observed ranking. Nevertheless, as shown in Fig. 9 the 2D Plot widget allowed users to model more precisely the preferences of each alternative, and was easier to use.

This experiment has shown that the 2D Plot Widget is easier to use and allows a better consistency for end-users. Nevertheless, we have to mention that the number of operable data is not consistent for all questions, as an example the satisfaction (perceived ease of use) conclusions are based on only 30 answers as the other conclusions are based on 37 answers.

To consolidate these results, the study must be conducted with more subjects. Another issue is that the subjects were students using computers easily. It could be interesting to compare two kinds of subjects: students not accustomed to using these Widgets and Information technologies students.

5 Conclusions

PCs are a central feature of AHP. They are simple to use; however, as the number of items to compare increases, so does the effort of conducting all comparisons and the probability of introducing inconsistencies increase as well. We presented an innovative approach to conduct PCs based on a UI widget that resembles an interactive data plot. It uses the transitivity property of a consistent comparison matrix to infer comparisons. Our hypothesis is that this new approach is more efficient (as it reduces the number of actions the user must conduct to compare all items), more effective (as it limits the sources of inconsistencies), and yields better user satisfaction.

Pairwise comparison tools frequently present each comparison independently. In contrast our widget presents multiple pairwise comparisons at once, visually suggesting how they relate to one another (especially via transitivity). Presenting comparisons this way may hinder (psychological) independence of comparisons. Moreover, the curve that connects handles may misguide users to believe that there is something between items (taking intermediate values). Studying the impact of these potential drawbacks is the focus of future work.

In the AHP methodology proposed by Saaty, inconsistencies are addressed thanks to the systematic pairwise comparisons. This way of capturing users' preferences ensures consistency of preferences as these preferences are processed at the deepest level of details. Nevertheless, this approach is very time consuming and we cannot guarantee that at the end of the process the end-user remembers his own choices done at the beginning of the process. Another approach to capture users' preferences is to ask them to directly evaluate the criteria and the alternatives on an ordinal scale. This approach has also shown its limits as it is very difficult for a decision-maker to proceed without any comparison. Thanks to the developed widget, our approach is a proposal to guarantee decision-makers' consistency by proposing global comparisons among criteria and alternatives and at the time the end-user saves time.

References

1. Saaty, T.L.: Fundamentals of Decision Making and Priority Theory, 1st edn. RWS Publications, Pittsburgh, PA (2000)
2. Triantaphyllou, E.: Multi-criteria Decision Making Methods: A Comparative Study, vol. 44. Springer, Boston (2000). https://doi.org/10.1007/978-1-4757-3157-6
3. Tanino, T.: Fuzzy preference orderings in group decision making. Fuzzy Sets Syst. **12**(2), 117–131 (1984). https://doi.org/10.1016/0165-0114(84)90032-0
4. Tversky, A.: Intransitivity of preferences. Psychol. Rev. **76**(1), 31–48 (1969). https://doi.org/10.1037/h0026750
5. Regenwetter, M., Dana, J., Davis-Stober, C.P.: Transitivity of preferences. Psychol. Rev. **118**(1), 42–56 (2011). https://doi.org/10.1037/a0021150
6. Benítez, J., Delgado-Galván, X., Izquierdo, J., Pérez-García, R.: Improving consistency in AHP decision-making processes. Appl. Math. Comput. **219**(5), 2432–2441 (2012). https://doi.org/10.1016/j.amc.2012.08.079
7. Saaty, T.L.: Super Decisions (2004). https://superdecisions.com/
8. Expert Choice: Expert Choice (2021). https://www.expertchoice.com

9. Siraj, S., Mikhailov, L., Keane, J.A.: PriEsT: an interactive decision support tool to estimate priorities from pairwise comparison judgments. Int. Trans. Oper. Res. **22**(2), 217–235 (2015). https://doi.org/10.1111/itor.12054
10. Kämpke, T., Radermacher, F.J., Wolf, P.: Supporting preference elicitation: the FAW preference elicitation tool. Decis. Support Syst. **9**(4), 381–391 (1993). https://doi.org/10.1016/0167-9236(93)90048-8
11. Abel, E., Galpin, I., Paton, N.W., Keane, J.A.: Pairwise comparisons or constrained optimization? A usability evaluation of techniques for eliciting decision priorities. Int. Trans. Oper. Res. November 2020. https://doi.org/10.1111/itor.12907
12. Millet, I.: The effectiveness of alternative preference elicitation methods in the analytic hierarchy process. J. Multi-criteria Decis. Anal. **6**(1), 41–51 (1997). https://doi.org/10.1002/(SICI)1099-1360(199701)6:1%3c41::AID-MCDA122%3e3.0.CO;2-D
13. Brunelli, M.: A survey of inconsistency indices for pairwise comparisons. Int. J. Gen. Syst. **47**(8), 751–771 (2018). https://doi.org/10.1080/03081079.2018.1523156
14. Saaty, T.L.: The Analytic Hierarchy Process: Planning, Priority Setting, Resource Allocation. McGraw-Hill, New York (1980)
15. Brunelli, M.: Introduction to the Analytic Hierarchy Process. Springer International Publishing, Cham (2015). https://doi.org/10.1007/978-3-319-12502-2
16. Brooke, J.: SUS - A quick and dirty usability scale. p. 8.
17. Sauro, J.: A Practical Guide to the System Usability Scale: Background, Benchmarks & Best Practices. CreateSpace Independent Publishing Platform, Denver (2011)

A Problem Structuring Multimethodology to Support a Post-graduation Course on Implant Dentistry

Silvana Marques Miranda Spyrides[1(✉)], Letícia Meinberg Pedrosa[1,2], Marcos Pereira Estellita Lins[2,3], and Clarice Guimarães Barros Martins[2]

[1] Department of Prosthesis and Dental Materials, School of Dentistry, UFRJ - Federal University of Rio de Janeiro, Rio de Janeiro, RJ, Brazil
silvanaspyrides@odonto.ufrj.br

[2] Department of Production Engineering, UFRJ – Federal University of Rio de Janeiro, Rio de Janeiro, RJ, Brazil

[3] Department of Production Engineering, CCET/UNIRIO - Federal University of State of Rio de Janeiro, Rio de Janeiro, RJ, Brazil

Abstract. The Federal University of Rio de Janeiro's extension course in Implant Dentistry is a pioneering successful initiative of the university's Dental School, that provides free access to implant services through an innovative modality of educational training. In recent years, however, the lack of public investments in universities has negatively affected these kinds of initiatives, forcing the institutions to adapt to maintain the supply of vacancies and the quality of education offeredQuery. It is essential, therefore, to evaluate the structures and management processes to enable both the continuity of teaching and training and the implant services to low-income people. In order to overcome these problems, we applied a multimethodology for structuring and proposing interventions called Complex Holographic Assessment of Paradoxical Problems (CHAP^2). This methodology, in contrast to those that isolate the parts of a problem and move quickly to its solutions, encompasses two levels: first, the perception of it as a complex problem and then the intervention on the problem to integrate the perspectives of the analysts and the agents involved. Concern the problem structuring, decision making and interventions. We identified and addressed the problems named "Disorganization and loss of medical records and administrative information", "Lack of data collection and consolidation of the cases treated in the course" and "Lack of performance indicators and goal setting". Concrete measures included the implementation of the database and respective indicators, the construction of the workflow in process maps and determining a new sliding-scale fee based on income for the dental implant service. The contribution of the application of CHAP^2 to the Extension Course in Implantology not only impacted the results on the management of the course, but also influenced its legacy, as it has engaged the agents involved and expanded their perspectives about the system.

Keywords: Dental implantation · Community-institutional relations · Health care · Management · Multimethodology

A. T. de Almeida and D. C. Morais (Eds.): INSID 2021, LNBIP 435, pp. 135–148, 2021.
https://doi.org/10.1007/978-3-030-91768-5_9

1 Introduction/Purpose

In Brazil, the vocation of the Universities is not restricted to Teaching and Scientific Research but encompasses University Extension as stated in the Federal Constitution, which prescribes the inseparability of these three pillars. University Extensions aims to provide a relationship between educational institutions and the society to which it belongs.

Public funding is the largest source of financial resources for any Brazilian Public University. National and international political, social, economic, and environmental crises, however, have imposed severe cuts in the budget of these universities. As a result, and extension activity has been seriously affected. The Extension Course in Implantology (ECI) at Faculty of Dentistry (FD) of the Federal University of Rio de Janeiro (UFRJ) is an example, as it is going through a crisis due to the lack of infrastructure, logistics, financial and human resources.

Studies aiming at enhancing the sustainability of the University Extension are then extremely relevant, as continuing education, like university extensions, allows for an improvement in the knowledge and competence of professional who ventures in it, such as, in this case the Dental Surgeon (Van Hoof and Meehan 2011).

Implantology is an area of Dentistry that provides rehabilitation for lost teeth through dental implants. It's an extremely important area, regarding the current oral condition in Brazil, where the average of missing teeth in the population aged between 65 and 74 is 25.4 teeth, amongst the 32 teeth in the permanent dentition (SB Brasil, 2010). However, procedures related to dental implants are not accessible to most of the Brazilian population due to the high costs involved.

Brazil is a country that presents important social and economic contrasts. Income inequality indicators are causally related to health indicators, disclosing income disparities as an important cause of poor health conditions (Szwarcwald et al. 1999). The integration between professional qualification and dental service to the poorest community has far-reaching social, cultural and economic benefits, regarding citizenship, quality of life and economic welfare, far beyond the private market profitability (Yoder 2006). Moreover, it promotes a process of reflection for the student, facilitating personal and professional development, and developing empathy, communication skills, and self-confidence (Modifi et al. 2003).

The Extension Course in Implantology (ECI) at the Faculty of Dentistry (FD) of the Federal University of Rio de Janeiro (UFRJ) was created in 2015 by the faculty of the Department of Prosthesis and Dental Materials (DPDM). It aims to integrate research, teaching and services, by training professionals in innovative technologies in clinical-surgical and prosthetic-laboratory implantology, made available through quality services at affordable prices. These dental treatments are targeted at the poorest population since the cost per implant for each patient is equivalent to about 21% of the median of the values currently charged in private clinics in the city of Rio de Janeiro.

The ECI is free of charge; however, the students must bear the costs of instruments and materials for individual patients. In consequence, after an initial consultation, the patient who fits and agrees to the available treatment must contribute with a complementary fee of R$ 300.00 (±US$ 77.00 in March of 2019) per dental procedure. The fee is used to face maintenance costs of undergraduate and specialization clinics, following the Rules of the FD of UFRJ.

However, nowadays the course faces some problems that hinders a good performance: the fee has become insufficient to meet the need for maintenance; the infrastructure does not favor the flow of information and the organization of work is poor.

To support the structuring of problems of the ECI and the search for solutions, a project was started to help to characterize its main issues, including the course structure, social service, financial resource, infrastructure, and management of the information.

Problem structuring methods apply to a productive system in its complexity, aiming to identify and model the problems and to intervene through formal methods to aid decision and management. In particular, the mapping of activities and their interrelationships provides a systemic approach for process management, which will allow the articulation of specific interventions. The literature presents some cases of application of management methodologies to dental services. Ahmed and Amagoh (2014) identifies all stages of services in a dental clinic and develops a plan to address bottlenecks, seeking to optimize and improve their services. Chapko et al. (1984) demonstrated positive effects when dental surgeons and assistants attended a workshop with management content, decision making, development of goals, programming, and communication in the dental office. Silva et al. (2019) stated that the mapping of processes allows an understanding of each process and collaborates in the identification of the hierarchy of processes, allowing the recognition of key processes. With this, there is the provision of a basis for decision making, which can be applied to dental clinics, assisting managers in understanding the routine of the clinic.

A metacognitive attitude contributes to the decision-making process of complex social problems. Through metacognition it is possible to provide mental detachment and develop a reinterpretation of the environment and relationships on your part, leading to a deepening of awareness and facilitating the decision-making process (Lins et al. 2021).

To support the continuity of the ECI as a free teaching channel for students and cheap, but qualified treatment for the society, the problem structuring methodology Complex Holographic Assessment of Paradoxical Problems (CHAP^2) was applied to provide an expanded view for the agents involved in the processes and to guide their necessary interventions on the identified problems (Lins e Antoun Netto 2018; Lins et al. 2021).

2 Methodology

We propose the use of a multimethodology for structuring the system's problems, where a problem structuring method named CHAP^2 plays a central role in integrating qualitative and quantitative methods.

Problem structuring is one of the stages of the decision-making process that aims to structure issues, problems, to later select methods and formulate models to deal with decision-making situations (Rosenhead and Mingers 2001). Problem structuring methods emerged in the late 1970s/early 1980s to support the formal mathematical approach (Ackoff 1978). They provide a broader, fundamental environment in the implementation and validation phases of modeling, particularly in situations where there is no clear agreement as to what the exact problem or its solution (Ackerman and Eden 2011). According to Pidd (2010), we consider the systems composed of four characteristics: limits that define the whole, components contained within the limits, processes that describe internal organization and behavior that emerges from the whole and not only from the components. Rosenhead and Mingers (2011) report a large number of methods, which present contrasting and convergent characteristics about products, processes, and complexity.

Multimethodology consists of using a combination of methods, and according to Mingers (2006) "refers in general to utilizing a plurality of methods or techniques, both qualitative and quantitative, within a real-world intervention". Howick and Ackerman (2011) report that "where the case reports the combination of two quantitative methods there is a tendency for the case to have an operational objective whereas combining two qualitative methods tend to concentrate on problems of a more strategic nature". Munro and Mingers (2002) noticed that the combination of qualitative and quantitative occurs in those interventions with more than two methods. Lins et al. (2019) used a multimethodological approach, combining concept maps with data mining and data envelopment analysis to assess the performance of municipal health in Brazil. A large number of authors approached applied health problems from qualitative (Rosenhead and Mingers 2001; Midgley 2000) or quantitative (Ozcan 2008; Sherman 1984) methods, to cite a few, however not using a formal multimethodology for an endogenous modeling of both approaches.

The present study uses the $CHAP^2$, which comprehends an interface that integrates a qualitative thematic mapping and several formal models that allow quantitative modeling. Awareness of social metacognitive processes is facilitated by cognitive maps and provides a context to help manage complex unstructured problems involving interactions between human, technological, organizational, and environmental components. This context supports the identification of specific problems to be faced using formal quantitative models to assist in decision-making in the later phases of $CHAP^2$ (Brinol and DeMaree 2012).

Complex social systems have paradoxes that need identification for proper management. The herein proposed dialogical and comprehensive view supports the disclosure of the differences between the perspectives of the agents involved and the potential conflicts, this being an important factor for choosing this methodology. The $CHAP^2$ proposes to facilitate the perception of organizational processes by the agents involved, favoring self-management and self-regulation of activities in the system; this includes the identification of paradoxes, which leads to a broadening of the organizational awareness (Laricchia 2015). Another important prerogative is that it is a multimethodology, which includes an interface between the qualitative context and the formal model with quantitative indicators and goals.

The application of $CHAP^2$ to the ECI was carried out in six phases (Fig. 1). In Phase I the objectives are a prior characterization of the system and the identification of a group of agents who develop activities in the system's processes. In the ECI, this stage developed through preliminary meetings with the head of the DPMD and the Course Coordinator, and an analysis of available documentation about the course, such as its notices and information published on the website of the FD of the UFRJ (Extension

Course in Implant Dentistry of the Dental School of the University of Rio de Janeiro 2019). The group consisted of the head of the DPMD, the course coordinator, two permanent teachers, and a substitute teacher.

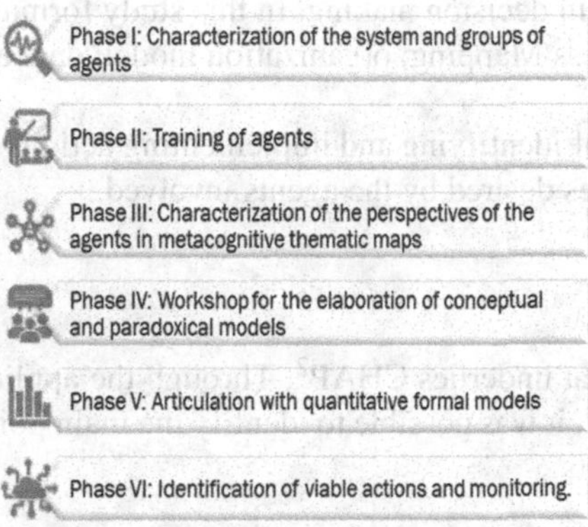

Fig. 1. Sequence of phases of CHAP2

Phase II aims at training/guiding the group of agents: it is carried out through seminars, group dynamics, or individual orientations, where the participants apply the management holographic metaphor. This consists in improving their skills to perceive the integration of its activities in the context of organizational processes, which promotes access to a broader perspective regarding the organization's main issues. In this work, guidance was carried out individually, due to the unavailability of common schedules amongst agents.

Phase III aimed at structuring the system through knowledge networked diagrams called concept or cognitive maps. The purpose of the cognitive map is the compilation and structuring the relevant factors to solve the problem (Pessoa et al. 2015). We interviewed the five agents and elaborated individual maps from their perspectives of the system, addressing problems, relevant issues, and rewarding results. Subsequently, we used a script previously designed, based on the following points: Definition of the context of the course; Course structure and operation; Main procedures; Positive points of the course; Improvement points; and future perspectives for the Course; leaving the interviewed agents free to add any information they considered important to the process. The individual maps were validated individually with each agent. Individual maps are analyzed comparatively, seeking to identify predominant themes and aggregate a thematic map. These will be used in the next phase for meta-cognitive analysis and identification of intervention opportunities. The consolidated Thematic Metacognitive Map was composed by five clusters: course structure, social service, financial resources, infrastructure, and information management. These maps intend to facilitate the metacognitive dialogue in the following workshop.

Phase IV consisted in a workshop to identify and prioritize problems and methods for interventions in the several clusters of the thematic map. Participants discussed and characterized the main problems. Afterwards, they proposed methods to direct interventions, composing the so-called conceptual model, which displays the convergences amongst the agents' perspectives. They also composed the paradoxical model, which displays the barriers to achieve the proposed solutions.

In Phase five, the interface between qualitative modeling and formal quantitative models is crossed. Formal models aim to provide process indicators and processes to support system's agents in decision making. In this study formal methods were applied through: workflow process Mapping; organization model; data and indicator treatment; and market research.

Phase VI consisted of identifying and implementing actions that are feasible for the interventions and changes desired by the agents involved.

3 Results

Systemic thinking is what underlies CHAP². Through the application of the method in the case study of the ECI, it was possible to identify the main points raised by the agents and outline improvements.

3.1 Problem Structuring

In Phase I, the characterization of the "real" system was obtained, which resulted in the Conceptual Map (Fig. 2), concerning the problems raised by the agents. The Conceptual Map exhibits the characteristics of the ECI, such as the functioning, the objective, the agents involved, profile of the patients attended, phases and characteristics of the treatment, and the interrelation of the course with other clinics in the DPDMs, with other departments, with community, and students.

The data from the individual maps were gathered into clusters that compose the Thematic Metacognitive Map (Fig. 3). The map is called metacognitive, not because of a structural property, but of a functional one, as it facilitates and promotes the use of intra and interpersonal intelligence.

3.2 Decision Making

From a systemic view and the analysis of the main aspects mentioned in Phase III, shown by the Thematic Metacognitive Map, some problems were identified and developed in the workshop. The discussions of problems within the five themes allowed the proposition of solutions and the identification of possible conflicting perspectives that may hinder well succeeded interventions. Table 1 shows the problems, propositions, and barriers to solutions for the five themes.

We prioritized some of the main problems identified, regarding expertise of the agents, as an object of modeling in Phase V. During the workshop, the participants chose the disorganization of information as the most critical point, including all issues directly or indirectly related to it. The gap between the fee charged to the patients and the current market value was also a relevant critical point.

The problems identified as critical also correlate to non-prioritized problems. Actually, structuring the problem and characterizing the real system brings a wide range of views that lead to solutions that consider the whole system. This emphasizes the need for a robust structuring of problems before solution proposals.

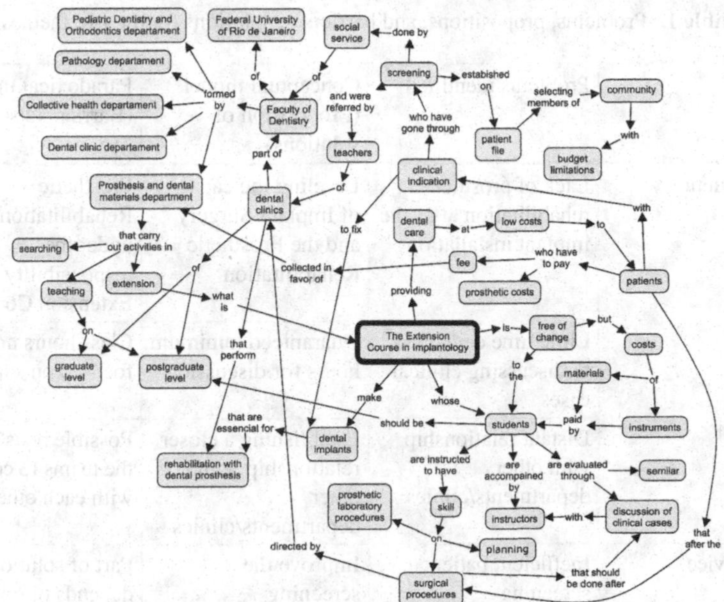

Fig. 2. Concept map - phase

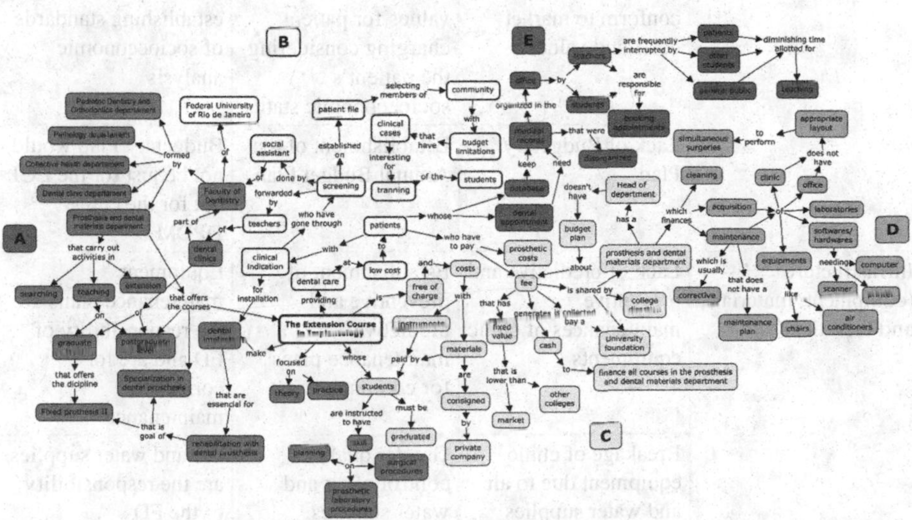

Fig. 3. Thematic metacognitive map. shows 5 clusters: course structure (A), Social service (B), Financial resources (C), Infrastructure, (D) Information management (E)

Table 1. Problems, propositions, and barriers to solutions for the five themes

Thematic	Problems identified	Conceptual model (Proposition of solutions)	Paradoxical model (Barriers to solutions)
Course structure	Lack of prosthetic rehabilitation after the implant installation	Leveling the capacity of Implant Surgery and the Prosthetic Rehabilitation	Prosthetic Rehabilitation is not under the responsibility of the Extension Course
	Little time dedicated to discussing clinical cases	Guaranteed minimum hours for discussion	Class hours normally focused on surgery
	Distant relationship with other departments/clinics	Establishing a closer relationship with other departments/clinics	Possible resistance of the teams to cooperate with each other
Social Service	Inefficient patient screening	Improve the screening	Part of solution depends of other departments/clinics
	Failure to monitor treatment from start to finish	Rehabilitation clinics should hold accountable explain treatment complete	Part of solution depends on the involvement of other departments/clinics
Financial	The fee doesn't conform to market costs and values	Establishment of new values for patient charging considering the patient's socioeconomic status	Difficulty in establishing standards of socioeconomic analysis
	Lack of Budgetary Plan	Establishment of the Annual Budget Plan	Budgetary Plan would not be just for the ECI but for the entire DPDM
Infrastructure (equipment, materials, and facilities)	Lack of predictive and preventive maintenances of clinic equipments	Establishment of preventive and predictive maintenance plans for equipments	Equipment maintenance plans are the responsibility of FD and are for corrective maintenance
	Breakage of clinic equipment due to air and water supplies	Greater quality control of air and water supplies arriving at DPDM	Air and water supplies are the responsibility of the FD

(*continued*)

Table 1. (*continued*)

Thematic	Problems identified	Conceptual model (Proposition of solutions)	Paradoxical model (Barriers to solutions)
	Consigned strategic equipments due to partnership termination	Analysis of strategic equipments options for acquisition	Possible lack of financial resources for immediate acquisition
	Lack of Information Technology equipment	Analysis and acquisition of Information Technology equipment	Possible lack of financial resources for immediate acquisition
	The layout of clinics, classrooms and other rooms of the DPDM hinders routine course activities	Development of a new layout to easy access from the clinic to other areas of the Department	Possible lack of financial resources for immediate rebuilding facilities
Information management	Disorganization and loss of important medical records and administrative information	Appointment of a qualified clerk to the secretary services, such as scheduling appointments and file organization	Hiring a clerk depends on public tender by University administration
	Wasting time of teaching staff for administration purposes	Designation of a qualified clerk for secretary services, such as scheduling appointments and file organization	Hiring a clerk depends on public tender by University administration
	Lack of data collection and consolidation of the cases treated in the course	Establishment of processes and controls for data collection and consolidation	Establishment of a culture of processes and controls
	Lack of performance indicators and goal setting	Data organization and establishment of key indicators for process analysis	Establishment of a culture of processes and controls

3.3 Intevention

It is important to emphasize that CHAP[2] stands out for making explicit the interface between the broad qualitative model in Phase IV and the formal models to support decision making in Phase V. The construction of conceptual and paradoxical models helped to characterize and prioritize issues and also to identify the variables that make up the indicators. Four main formal methods were applied: process mapping; organizational model; database and indicators treatment; and market research.

The process mapping and the organizational model allowed a comprehensive understanding of the personnel, the material requirements and of how the implemented activities related to each other in order to formulate and prioritize interventions. The database and indicators, contrastingly, delivered a quantitative model. The last tool, market research, focused on the prioritized financial problem and used both qualitative and quantitative methods.

Process Mapping. Defined as a set of activities carried out in a logical sequence to produce a product or service that has value for a group of customers (Hammer and Champy 1994). The mapping of processes collaborates in the strategic organization and forms a basis for decision making (Silva et al. 2019).

From the analysis of the process, we observed how students and teachers were involved in auxiliary activities, such as scheduling appointments and separating medical records. This doesn't add value to the objective of the course, which is to enable students to plan and execute surgical procedures concerning implant installation.

Thus, a process proposal was elaborated (Fig. 4) where these auxiliary activities would be concentrated and carried out by a qualified clerk. It was difficult to hire a qualified clerk by the University, however, we could hire a trainee, who became responsible for collecting resources generated by the ECI. The centralization of the activities of control of the archives avoided the loss of information and medical records and also keep them organized.

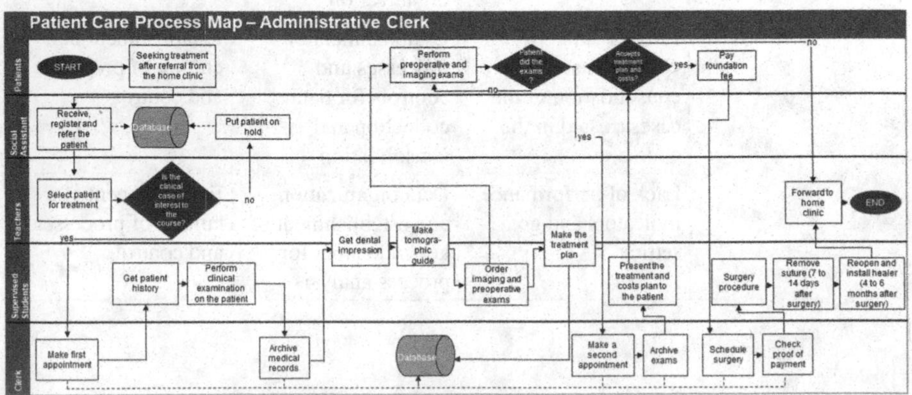

Fig. 4. Proposed process map for patient care

Organization Model. Based on the need for greater reliability of the data available for the ECI, we proposed a documental organization in a sequence of activities, which addressed physical and virtual data. The physical file was alphabetically organized, to separate ongoing from completed cases. For the virtual archive, we developed an electronic spreadsheet with all the information concerning patients, as an integrated database.

Treatment of Data and Indicators. The documentation process allowed to organize data into bases for the establishment of indicators. We intended to formulate performance indicators to carry out periodic measurements and the evaluation of the results obtained by organizations and propose short, medium, and long-term goals.

We developed an electronic spreadsheet where all patient information was added, such as: name, gender, age, address, contacts, marital status, profession, level of education, and monthly income was integrated to automatically generate social indicators. This information will support decisions about alternative proposals for fixing the charged fee.

Regarding implant services, this spreadsheet records information about treatment, such as: clinic of origin, appointment dates, surgery dates, number of implants installed, types of implants, and amount paid, in addition to important observations. Therefore, besides patient social information, the database generates indicators regarding course performance (number of surgeries and implants installed per student and per year) and financial performance. This intends to respond to the following concerns reported in the workshop:

- Follow-Up of the treatment from start to finish.
- Coping with the delay and lack of prosthetic rehabilitation that happens in some cases after installing the implants. In this sense, the number of patients originating from each clinic can be monitored, tracing back deviations, and ensuring complete rehabilitation: implant and prosthetic.
- Lack of a budget plan. With the monitoring of historical data on installed implants and financial revenue, it is possible to make budget plans, thus allocating the resources for maintenance and investments.

Market Research. The target clientele of an educational institution is the low-income population. However, there is also demand from some higher-income patients, given the institution's high quality treatments and the much lower prices. The priority problem in the financial theme was the fee charged on patients, which was out of date. Regardless of the patient's purchasing power, the fee charged was US$ 77.00 per implant made.

After analyzing the researched data, it was decided to continue charging the patient of the course a fee per implant made, but then the patient's socioeconomic situation would be taken into consideration when defining the amount to be charged. One of the reasons for those criteria was to cover the basic maintenance costs, as that would ensure the continuity of the provision of high-level treatments to low-income patients and the quality education to students. Actually, the surplus from the higher fees on wealthier patients can be converted into more investments in the clinics of DPDM and, in the future, can finance treatments free of charge to very low-income patients.

To establish a price ceiling for the fee, we first carried out two surveys, one at educational institutions and another at private clinics, both inquiring into the offered price for the same implant installations. The maximum fee was established as the minimum average for implant procedure values amongst the surveyed clinics, which was US$ 289.00. Then we assumed that fees should vary inside the range US$ 77.00 to US$ 289.00.

To define the fees to be charged within that range, we used information on household consumption expenses from the Family Budget Survey 2008–2009 of the Brazilian Institute of Geography and Statistics (IBGE 2010). Considering the seven classes of monthly income, we used the percentage of health expenses in 2009 and the upper limit of the monthly income class in 2019 for the estimates of fees to be charged, shown in Table 2.

Table 2. Expenses on health per income class and a proposed fee per implant

Monthly income (US$) (A)	Minimum wages (B)	% of health expenses (C)	Health expenses maximum value in (A) × (C)	Fee amount in US$
Up to 512,30	Up to 2	6,8	34,83	77,00
512,30–768,46	2–3	6,5	50,02	77,00
768,46–1.536,92	3–6	6,6	101,43	101,38
1.536,92–2.561,53	6–10	6,1	156,25	156,56
2.561,53–3.842,30	10–15	6,0	230,53	231,00
3.842,30–6.403,83	15–25	6,6	422,65	289,00
More of 6.403,83	More of 25	6,3	403,44	289,00

Thereafter, it was proposed that the patient's fee will be established based on a proofed patient's monthly income submitted at an interview with the unit's social worker.

Phase VI of the CHAP² method is characterized by the implementation of feasible actions for the changes and transformations desired by the agents involved. The methodology proposes both external and self-regulation when implementing viable actions.

4 Conclusions

The contribution of the application of CHAP² to the ECI wasn't limited to the results on the management of the course, but must be seen as an actual engagement of the agents involved in the entire process and then in the consequent expansion of their perspectives about the system.

This approach motivates agents to address problems besides those prioritized so far, taking advantage of their new perspective displayed in the metacognitive maps. The greatest contribution of CHAP[2] is this expansion of consciousness and perspective, since it facilitates changes.

In this work, the following interventions were done, in attendance to the changes requested by the agents involved in the selected problems:

- the designation of a clerk for the course, whose work resulted in the organization of the manual files.
- the implementation of the virtual files and respective indicators, in addition to the others bureaucratic functions.

In this way, we addressed the problems identified as "Disorganization and loss of medical records and administrative information", "Lack of data collection and consolidation of the cases treated in the course" and "Lack of performance indicators and goal setting".

With regard to the other selected problem "Out-of-date charge for patients", the administrative team accepted the implementation of proposed fees per implant to be charged by income classes as a solution, being up to the unit's social worker to define the income class to which the patient belongs and the fee to be charged.

Some of the problems that arouse in Phase IV, although they were not prioritized, were resolved through individual initiatives even while applying the methodology. Among them, we can mention the purchase of peripheral dental equipment and the change of the clinic layout to facilitate the routine activities of the course.

Application of the CHAP[2] methodology to other Extension Courses in Public Universities, or to dental public services, in general, can contribute to improving performance and sustainability, given their relevance for society and scarce financial resources. As a matter of fact, the methodology is already under application for supporting diagnosis and problem structuring in a municipality dental service in Rio de Janeiro, for prosthetic services.

Acknowledgments. Supported by the Brazilian agency: CNPq (Grant#303346/2017-5).

References

Ackerman, F., Eden, C.: Strategic management of stakeholders. Theory Pract. **44**,179–196 (2011)

Ackoff, R.L.: The Art of Problem Solving. John Wiley and sons, Chichester (1978)

Ahmed, S., Amogoh, F.: Process analysis and capacity utilization in a dental clinic in Kazakhstan. Compet. Rev. **24**, 347–356 (2014)

Brinol, P., DeMaree, K.G.: Social Meta-Cognition. In: Frontiers of Social Psychology. Psychology Press, Hove (2012)

Chapko, M.K., Milgrom, P., Bergner, M., Conrad, D., Skalabrin, N.: The effects of continuing education in dental practice management. J. Dent. Educ. **48**, 659–664 (1984)

Extension Course in Implant Dentistry of the Dental School of the Federal University of Rio de Janeiro https://xn--extenso-2wa.ufrj.br/index.php/67-acoes-de-extensao/cursos-deextensao/saude/322qualificacao-profissional-implantodontia. Accessed 4 Mar 2019

Hammer, M., Champy, J.: Reengineering the Corporation. Harper Brothers Publishers, New York (1994)

Howick, S, Ackerman, F.: Mixing or methods in practice: past, present and future directions. Eur. J. Oper. Res. **215**, 503–511 (2011)

Instituto Brasileiro de Geografia e Estatística: Pesquisa de Orçamentos Familiares 2008–2009 (2010). https://www.ibge.gov.br/. Accessed 3 May 2019

Laricchia, C.R.: Estruturação de Problemas Complexos na Agricultura Familiar: CHAP2 e Pesquisa-Ação. Dissertation, University Federal of Rio de Janeiro (2015)

Lins, M.P.E., Netto, S.O., Lobo, M.S.: Multimethodology applied to the evaluation of healthcare. J. Health Care Manag. Sci. **22**, 197–214 (2019)

Lins, M.P.E., Antoun Netto, S.O.: Estruturação de Problemas Sociais Complexos - Teoria da mente, mapas metacognitivos e modelos de apoio à decisão. Editora Interciência, Rio de Janeiro (2018)

Lins, M.P.E., Pamplona, L., Lins, A.E., Lyra, K.: Metacognitive attitude for decision making at a university hospital. Int. Trans. Oper. Res. 1–21 (2021)

Midgley, G.: Systemic Intervention – Philosophy, Methdodology and Practice, Kluwer Academic, New York (2000)

Mingers, J.: Realising Systems Thinking: Knowledge and Action in Management Science. Springer, Boston (2006).https://doi.org/10.1007/0-387-29841-X

Mofidi, M., Strauss, R., Pitner, L.L., Sandler, E.S.: Dental student's reflections on their community-based experiences: the use of critical incidents. J. Dent. Educ. **67**, 515–523 (2003)

Munro, I., Mingers, J.: The use of multimethodology in practice - results of a survey of practitioners. J. Oper. Res. Soc. **59**(4), 369–378 (2002)

Ozcan, Y.A.: Health Care Benchmarking and Performance Evaluation: An Assessment Using Data Envelopment Analysis (DEA). International Series in Operations Research and Management Science. Springer, Berlin (2008). https://doi.org/10.1007/978-1-4899-7472-3

Pessoa, L.A.M., Lins, M.P.E., Silva, A.C.M., Fiszman, R.: Integrating soft and hard operational research to improve surgical centre management at a university hospital. Eur. J. Oper. Res. **245**, 851–861 (2015)

Pidd, M.: Tools for Thinking: Modelling in Management Science. John Wiley and Sons, New York (2010)

Rosenhead, J., Rational, M.J.: Analysis for a Problematic World: Problem Structuring Methods for Complexity, Uncertainty and Conflict. Wiley and Sons, Chichester (2001)

SB BRAZIL National Research on Oral Health: main results (2010). http://bvsms.saude.gov.br/bvs/publicacoes/pesquisa_nacional_saude_bucal.pdf. Accessed 2 Nov 2019

Silva, G.N., Kawamoto, L.T., Jr., Nery, L.A.S.S., Kawamoto, W.O., Sanchini, P.A.: Process mapping in a dental clinic. RIES **8**, 71–99 (2019)

Sherman, H.D.: Hospital efficiency measurement and evaluation. Med Care **22**(10), 922–939 (1984)

Szwarcwald, C.L., et al.: Desigualdade de renda e situação de saúde: o caso do Rio de Janeiro. Cad. Saúde Pública. **15**, 15–28 (1999)

Van Hoof, T.J., Meehan, T.P.: Integrating essential components of quality improvement into a new paradigm for continuing education. J. Contin. Educ. Health **31**, 207–214 (2011)

Yoder, K.M.: A framework for service-learning in dental education. J. Dent. Educ. **70**, 115–123 (2006)

Author Index

Printed in the United States
by Baker & Taylor Publisher Services